AF589384

Resource Geography of Goa

THE AUTHOR

Dr Dadapir Moulasab Jakati completed his Post-Graduation in Geography from Karnataka University, Dharwad and M.Sc. in Ecology and Environment from Sikkim Manipal University, Gangtok, Sikkim. He obtained his M. Phil in Resource Geography and Doctoral degree in Resource and Marketing Geography. Presently, he is an Associate Professor of Geography and has been teaching in the Under-Graduate Department of Geography, St. Xavier's College, Mapusa, Bardez, Goa. He is engaged in research in the specializations of Resource Geography, and Marketing Geography. His research papers are published in the known professional journals. He is also presented the research papers in the both National and International Conferences in India. Beside these, he is Promoter of Environmental Education of IIEE., New Delhi. Life member of Geographers Association Goa. and Ph.D. Guide/ Research Supervisor of J.J.T. University, Rajasthan. He was elected as Member to the University Court of Goa University, Goa.

Resource Geography of Goa

– Author –

Dadapir Moulasab Jakati

2019

Scholars World

A Division of

Astral International Pvt. Ltd.

New Delhi-110 002

ISBN: 9789390384501 (Hardbound)

Published by : **Scholars World**

A Division of

Astral International Pvt. Ltd.

– ISO 9001:2015 Certified Company –

4736/23, Ansari Road, Darya Ganj
New Delhi-110 002
Ph. 011-4354 9197, 2327 8134

E-mail: info@astralint.com

Website: www.astralint.com

Digitally Printed at : **Replika Press Pvt. Ltd.**

Foreword

Goa, though a tiny state as indicated on the map of India, attracts a large number of people from all over the world. Known for its rich resources like mountains, rivers, beaches, it is also popular for its rice-curyy and fish.

Resource of Geography of Goa is an excellent initiative to present to the students the various resources of Goa. Every Goan should be proud of its bio-diversity with rich fauna and flora. This book reveals the treasure of Goa's resources. The future generation will get a picture of what Goa is and keep the reminiscence alive in the minds of every Goan.

The author has done an intense study to portray the picture Goa's resources by paying due attention to its past as well as the present development of Goa's resources. The author tried to cover and give due recognition to every aspect of Goa's resources. It has minutely focused on the climatic conditions, demographic structure and also touching upon Goa's population.

This book presents ion details the economic characteristics of Goa which include agriculture, mining, industries, power sector, banking, tourism and fishery and not excluding the growth of communication ion Goa.

OFFG. PRINCIPAL
ST. XAVIER'S COLLEGE
MAPUSA - GOA.

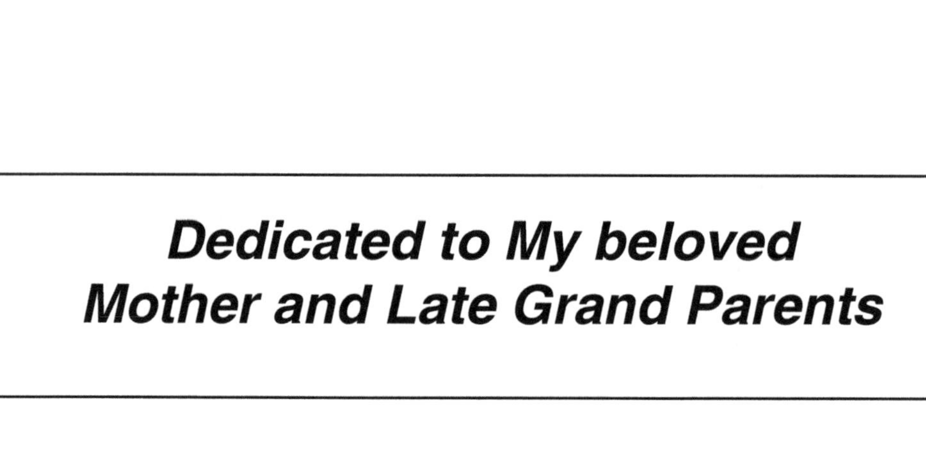

Dedicated to My beloved Mother and Late Grand Parents

Preface

Goa is one of the tiny states of India. It has ecstatic beauty that attracts the tourists from all over the country as well as world, and therefore, popularly known as a tourist state. It is well endowed with natural resources, i.e., minerals, forests, fisheries as well as agriculture. The mining activities are predominant, and also a key factor for industrialization. Goa is also known for popular drink cashew fenny. As a result, these aspects are determining the economy of Goa. Because there was a colonial rule by the Portuguese for a period of 450 years in Goa, still the impact of Portuguese culture can be seen all over Goa. The movement of people from other states brought the diversity in the cultural environment of Goa. Therefore, an attempt has been made to understand the geographical personality of the study area, in terms of its physiography, geology, climate, drainage, natural vegetation and soil, etc. in the form of physical characteristics.

Geography has become an extremely varied, versatile and high empirical discipline, and its area of study has grown a great deal since its inception. Harvey (1969) opines that geography is concerned with the description and explanation of the areal differentiation on the earth's surface. Growing human population is responsible for increasing demand for resources. The available resources are scarce in nature, on the one hand, and the population is growing rapidly on the other. Hence, it is the need of the hour to adopt the optimum utilization concept for effective conservation and management of resources.

Resource Geography of Goa has been taught at under graduate level prescribed by Goa University. The textbook will provide relevant information to the students about Goa mainly consists of History, administrative divisions, physical resources, economic, demographic and settlement characteristics of Goa etc. The content has been supported with illustration, graphs tables etc.

Thanks to our publisher ASTRAL International Pvt. Ltd., New Delhi for not only providing outstanding technical support but also for suggesting intricate modifications that would have never occurred to us.

Dr Dadapir Moulasab Jakati

Acknowledgement

At the outset I sincerely express my deep sense of gratitude to the almighty who has given me strength, patience and enlightenment in completion of the research work.

It is my proud privilege and great honor to express my deep sense of gratitude to my Research Guide and Supervisor Dr. Aravind A. Mulimani, Professor and Chairman, Department of Studies in Geography, Karnatak University, Dharwad, for his able guidance, unstinted encouragement, inspiring discussion and great support throughout this work.

Indeed I am grateful to the President and Secretary of the Diocesan Society of Education Management, Panaji, Goa. Rev Fr. Dr. Jeronimo Disilva, Rev Fr. Dr. Walter D'Sa , Principal of St. Xavier's college for their encouragement and cooperation throughout my research.

My deep sense of gratitude to Shri. Bhaskar G. Nayak Ex Director of Higher Education, Government of Goa. and Late Rev Canon Antimo Gomes, Ex Presedent of D.S.E., for their blessings, encouragement and great support in my academic career.

I must also express my sincere thanks to all the officials of Directorate of Fisheries, Government of Goa; Directorate of Census Operations; Directorate of Planning and Statistics Evaluation; Department of Industries and Mines; Department of Tourism; Goa Mineral Ore Exporters Association; Fishery Survey of India, Mormugao Indian Council of Agricultural Research

Complex, Old Goa; Marine Products Exporters Development Association (MPEDA); Exporters Inspection Agency CIA; Karnatak University Library Staff; Goa University; National Institute of Oceanography, Donapaula, Goa; St. Xaviers College Library.

I personally extend my sincere thanks to Dr. Subramanian, Dr. Mohanta, Scientists, Fishery Section, Indian Council of Agricultural Research, Old Goa, for their valuable suggestions and Shri Raghuram, Librarian, ICAR, for necessary help and cooperation.

My special thanks are due to my all the colleagues and friends of St. Xavier's college, Mapusa, Goa for their best wishes and blessings.

I shall be failing in my duty if I don't express heartfelt gratitude to my father Moulasab Jakati, mother Late Khairunisa Jakati, late Grandparents, uncles for their blessing, untiring support.

I am ever grateful to my beloved wife Farzana who stood by me in all times and helped me a lot in completion of my research work. I also acknowledge my lovely daughters Ms. Ruhinaaz and Ms. Rida for their loving support.

I sincerely thank Dr. Sawant, Vice-Principal of S.P. Chowgule College, Dr. R.B. Patil, M.E.S. College for their valuable co-operation and help. I am very thankful to Ms. Sumata and Mr. Subodh, Post Graduate students, Geoinformatic Centre, Goa, for impressive, cartographic work,

Dr Dadapir Moulasab Jakati

Contents

	Foreword	*v*
	Preface	*vii*
	Acknowledgement	*ix*
1.	**Geo–Look of the Study Area**	**1**
2.	**Demography – Population Structure of Goa**	**25**
3.	**Settlement Characteristics of Goa**	**39**
4.	**Economic Characteristics of Goa**	**42**
5.	**Fishery Resources of Goa**	**74**
	Bibliography	**87**

Chapter-1
Geo–Look of the Study Area

INTRODUCTION

Goa is one of the tiny states of India. It has ecstatic beauty that attracts the tourists from all over the country as well as world, and therefore, popularly known as a tourist state. It is well endowed with natural resources, i.e., minerals, forests, fisheries as well as agriculture. The mining activities are predominant, and also a key factor for industrialization. Goa is also known for popular drink cashew fenny. As a result, these aspects are determining the economy of Goa. Because there was a colonial rule by the Portuguese for a period of 450 years in Goa, still the impact of Portuguese culture can be seen all over Goa. The movement of people from other states brought the diversity in the cultural environment of Goa. Therefore, an attempt has been made to understand the geographical personality of the study area, in terms of its physiography, geology, climate, drainage, natural vegetation and soil, etc. in the form of physical characteristics.

The demographic characteristics are discussed in length. The settlement characteristics are the main considerations in the present study area, taking the population into account. The economic characteristics which form sound base for the economy of Goa. The transportation is the prime indicator not only to facilitate the tourism mining and industrial activities but also fishery activities. Therefore, these aspects form the basis for the regional development of Goa. The present study has been carried out only to focus on fishery resource and marketing development of Goa. The spatial planning strategy is required for the further development. Hence the geographical personality is essential to prepare a plan for future development.

HISTORICAL BACKGROUND

Before the advent of the Portuguese, early in the 16th century, the state covered very extensive areas, which extended towards the northern part of the Sindhdurg district of Maharashtra, then known as Kudal and Rajpur Mahals upto the river Kharepatan. The southern limit extended towards Ankola and comprised the ancient Mahals of Supa, Halyal and Karwar, now forming part of the North Canara district of Karnataka state. Towards the east it is covered by a larger portion of Belgaum district of Karnataka. Claude Alwares (2002).

The people of Goa are considered to be the descendents of Dravidians, who were natives of India. Dravids were gradually invaded by the Aryan Morch from the north around 1500 B.C. The Aryans after settling in western India established and named their state Konkan. Thus Goa was part of Konkan and gradually became an important part of the ancient and medieval traders like Phoenicians, the Persians, the Arabs, the Sumerians, the Greek and Romans.

Goa was under the control of several ruling dynasties that ruled. Goa can be seen in its diverse culture. The state was part of the Mauryan Empire under Ashoka, who reigned from 273-236 B.C. The Shathvahanas acquired Goa in the 2nd century B.C. After that the control went to the western Kshtrapas who ruled from 150 A.D. After that the Bhojas came to power and made Chandor as their capital, which is now known as Chandor. The Bhojas continued to rule over Goa from their capital at Chandrapur modern Chandor upto the 7th century A.D. The Rashtrakutas controlled Goa from 753 A.D. to 973 A.D.

The Kadambas exercised their power over Goa from 1008 A.D. to 1300 A.D. Goa became the maritime power of India during their rule. They built their capital at Gopak pattana. Ruled from there until their fall. It is now called Goa velha, which is a few kilometers to the south-west of old Goa. The Kadambas of Goa seem to have become independent by the early part of the 14th century A.D., when Yadavas were defeated at the hands of the Delhi sultans. Goa became a part of the Vijayanagara kingdom by the 14th century. In 1469, Goa passed under the Bahamanis sultan of Gulbarga when

Mohammed (1463-1482) conquered the Konkan area. With the break-up of the Bahamanis dynasty, it became a part of the kingdom of the Adilshahis of Bijapur in 1488.

In 1510, Alfonzo de Albuquerque, a Portuguese general, after a futile attempt at holding the city of old Goa, succeeded in driving out of the forces of Ismail Adil Shah (1510-1534), the sultan of Bijapur. By the middle of the 16th century, the Portuguese successfully established their rule in four talukas which are known as "velhas conquistas". By the end of the 18th century, the Portuguese were successful in annexing the remaining seven talukas of Goa known as "novas conquistas".

The 20th century was noted for blood shed and political uprising in Goa. The liberalization of Dadar and Nagar Haveli in 1954, gave a future filip to Goans' freedom movement. The government of India tried again and again to persuade the Portuguese government to withdraw peacefully. But the latter did not respond favourably. This adamant attitude of Portuguese ultimately forced the Government of India to send an army to liberate Goa. This historical event which marked the end of about 450 years of Portuguese rule took place on December 19, 1961 and thus Goa was liberated, and became a part and parcel of India on that day. Goa was a district of the union territory of Goa, Daman and Diu until May 30, 1987 when Goa attained the status of statehood.

LOCATION

Goa is a tiny emerald land situated on the west coast of Indian peninsular. It is located between 14? 53? 54? North latitudes to 15? 48? 00? North latitudes and 73? 40? 33? East longitudes to 74? 20? 13? East longitudes and is 1.022 meters above the mean sea level. It has an area of 3702 square kilometers with population of 13,47,668 as per 2001 census. The population of North Goa district was 7,58,573 and that of South Goa district was 5,89,095. As per the census, the density of population is 364 persons per sq.km. which is higher than that of all India level, and the neighbouring states of Karnataka and Maharashtra.

GOA – LOCATION

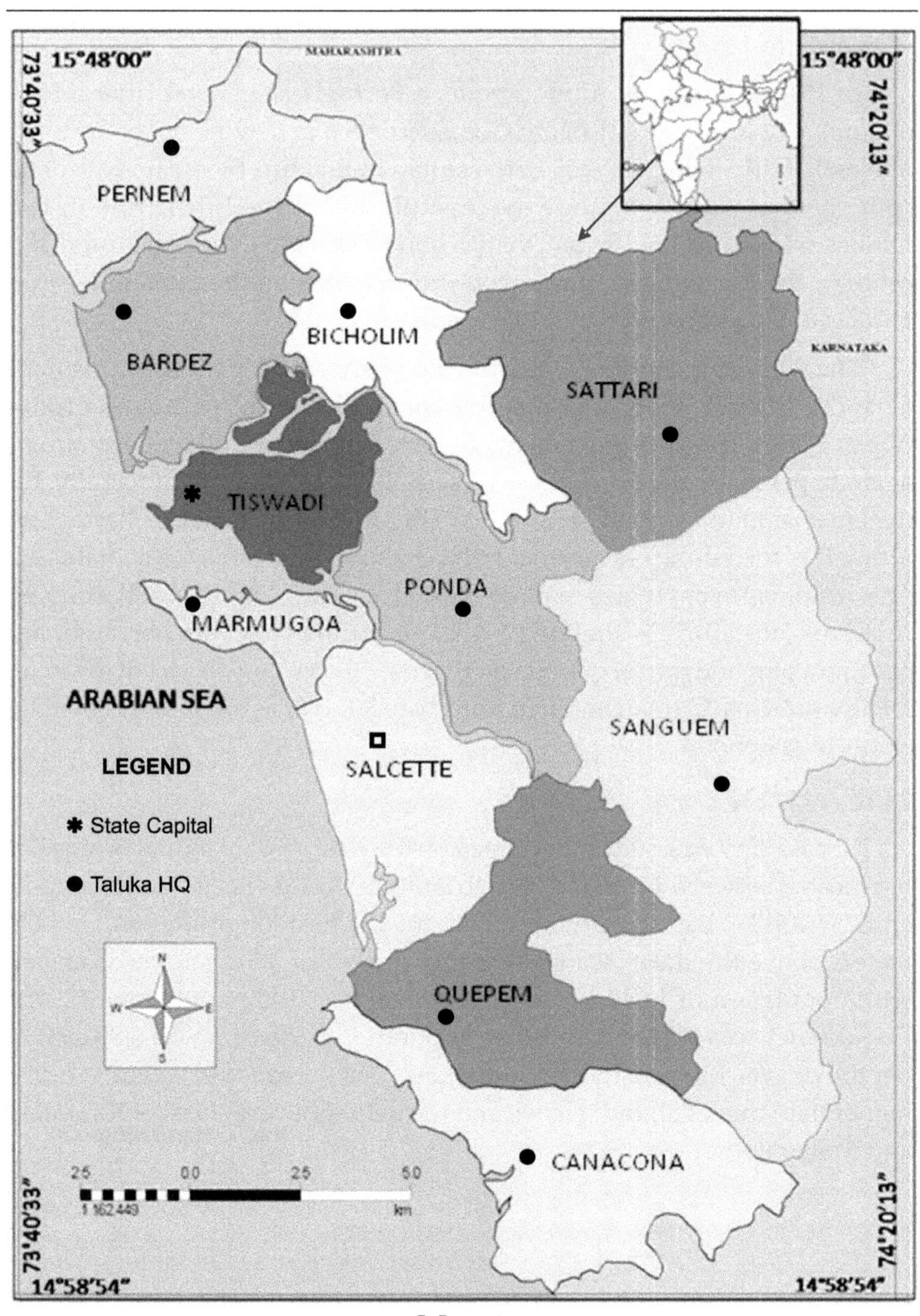

Map 1

Table-.1 : The Study Area at a Glance

Sl. No.	Taluka	Area in Sq.kms.	No. of Villages	No. of Census Towns	Population	Density /Sq.km.	Sex Ratio per 1000	Marine Fishing Village	Inland Fishing Village	Coastal Length in kms.
1.	Pernem	251.69	26	2	77999	286	943	4	-	8
2.	Bardez	263.97	33	11	227695	863	956	5	7	20
3.	Tiswadi	213.57	26	5	160091	750	969	2	10	17
4.	Bicholim	238.80	22	4	90734	380	943	-	-	-
5.	Sattari	489.46	78	1	58613	120	962	-	-	-
6.	Ponda	292.78	28	4	149441	510	937	-	6	-
7.	Marmugao	109.1	17	3	144949	1328	893	8	2	26
8.	Salcete	262.03	34	9	262035	895	1019	6	2	16
9.	Quepem	318.03	36	2	74034	233	973	-	-	1
10.	Sanguem	836.8	51	2	64080	77	967	-	-	-
11.	Canacona	352.1	8	1	43997	125	975	8	3	6
	Total	3702	359	44	1347668	364	958	33	30	104

Source: Economic Survey (2008-09), Directorate of Fisheries (2009-10)

The study area has an average density of 364 persons are living per sq.km area, sex ratio i.e. 958 females for every 1000 males which is quite low in the region. There are 33 marine fishing villages with fish landing centres as well as inland fishing villages are located in the internal parts comprising Bardez, Tiswadi, Ponda talukas. Coastal length has been estimated at 104 kilometers, out of which the North Goa has 45 kilometers while South Goa is having 59 kilometers largely benefiting the fishermen community.

The study area is surrounded by Maharashtra, Karnataka and Arabia Sea on its borders. The North Goa shares its boundary with the Savantwadi taluk of Sindhdurg district of Maharashtra state. The North of Terekhol lies between Goa and includes the Terekhol fort across the mouth of Terekhol river. In the south, North Canara (Karwar) district of Karnataka state and in the east it is covered by larger portion of Belgaum district of Karnataka. It has a coast of 104 kms long on the western side.

By administrative point of view, the state has been divided into two districts North Goa and South Goa. The North Goa comprises six taluks with an area of 1736 sq.kms. such as Perhem, Bardez, Bicholim, Sattari, Tiswadi and Ponda taluk, whereas, the South Goa comprises 5 taluks with an area of 1966 sq.kms., namely, Mormugoa, Salcette, Quepem, Sanguem and Canacona.

The study area has 359 villages, out of which 347 are inhabited villages (revenue villages), and the remaining uninhabited villages are submerged revenue villages. There are 189 village panchayats, known as local self government bodies, taking care of villages. For development works and to implement various projects, the state has been divided into 11 blocks (BDO's).

GEOLOGY

A substantial part of the land mass of the State of Goa consists of pre-Cambrian rocks like quartz. Scricites, schist, metavolvnics, with layers of granite and basalt wedged into them. In fact, Goa forms part of the basaltic outflows of the lava of the Deccan plateau. On that account, it has land forms consisting of flat topped peaks with terraced flanks and opening valleys with thin sides rising more in the form of steps than as smooth slopes. The topography of the basaltic formations is attributed to weathering and water erosion, giving rise to residual Hill, characteristics of rounded summits and minor mounds.

A small part of the area to its north is covered by the upper crateceous and lowereocone rocks of the Decean trap which have a thin cap of porous laterite stone formation. These rocks have undergone a process of laterisation in varying degrees all over the state especially along the coastal area, where they occupy a depth of fifty to seventy five (50-75 meters) from the surface. The process is attributed to the moist tropical climate of the region with its seasonal change of rain sun and cold. Such laterite caps also codes extensively the upper Sahyadris and the medium and low level plateaus stretch below them, containing the mineral bearing pink phylliterocks.

There are also in this region some out crops of rocks of older formations, like metamorphic schists and shales of a lesser scale. These rocks consists of subrounded to rounded cobbles, pebbles and boulders of a variety of quartzites and gveissic granite cemented in a greywacke motrix. Embeded in these formations, at some places, are the dolerite dykes. The literate is the latest formation in this area, with a thickness, of a mantle extending up to fifteen meters, depending upon the proximity to the sea coast, the nature and chemical composition of the original rock and the topography of the region.

More important than the above are the more recent alluvial formations that are spread all over the basins along the causes of rivers and the costal plains. Here are lands reclaimed from the river side and full of silt washed down their banks, known as the khazans, built up laboriously over thousands of years of meticulous planning, with sluice- gates to regulate the flow of water in and out of them, bringing in a fresh supply of nutrients, with much of its loamy land lying below sea level. These lands are purportedly the most fertile in the state, though threatened with irreparable damage due to recent developments. In a class like them are the sandy deposits along the coastline formed out of the debris of sand dunes that are swept away by the wind and the sea and by its interactions with the rivers, and tidal flows.

PHYSIOGRAPHY

Physiology is one of the major aspects to deal with the personality of the study region. Goa being part of the west coast region of India, many of its physical features are common to the neighbouring regions of Karnataka and Maharashtra states as well as other parts of the country. But there are some other characteristics of its own which are typical of this state, and lend its varied scenery a distinctive charm. Goa is basically divided into three major physiographic divisions, namely-

1. The Sahyadries – the mountainous region / (Western Ghats) in the east.
2. The Plateaus – the central portion of Goa plains.
3. The Coastal plains – River basins formed (formed by all the deposits in them and the coastal plains laced with sands near the sea level).

a) The Sahyadris – the mountainous region / Western Ghat in the east

The mountainous region of the Sahyadris that lies in the state has as an area about 600 sq.kms. and an average elevation of around 800 meters above the sea level. Its crestline assumes the shape of a mountainous arc of as escarpment which is about a 125 Kilometers long (Hundred and Twenty Five Kilometers in length). Viewed from the plains of Goa, it appears like a rampart, with peaks standing sentinel at various points of it connected with saddles of rock below and enveloped in a hazy azure blue with mists dominating over it, especially during the rains.

The watersheds that form in their midst act as sources of water for most of rivers that flow down them through Goan territory. The face of their escarpment is cut through at several places by upgraded streams that blossom into waterfalls.

The most well-known of them are the Harvalem and Dudhsagar waterfalls that descend down the green – clad mountains. Harvalem is more modest in the splendor of its flow, from the heights of the Ghats on Goas northern hindward side, on a lower altitude than the Dudhsagar falls, in the midst of rich flora and caves of yore with traditions of dwelling by the pandavas during their sojourns in the forest as depicted in the Mahabharata. Dudhsagar literally ocean of milk is a more majestic sight to behold, especially during the monsoons as the fulsome flow of swirling water cascades down in magnificent billows from the lofty peaks of Ghats through ever green luxuriant vegetation. The range of mountains along the outer periphery of the half moon formation that Goa becomes which has isolated peaks on them at several places. The most important of them are as follows: In the north, the Sonsogodd which attains an altitude of 3,827 feet above sea level. Katlenchimauli 3633 feet high, Vagheri Zormen 3500 feet high, Morlemgodd 3,400 feet high, all of them being in sattaritalaka. On the eastern side and the western side there are Siddhnath in Ponda, Chandranath in Paroda, Consid at Ashtragar and Dadhsagar at Latambarcem, which have lower altitudes slightly less than 3000 feet.

b) The Plateaus

The plateaus constitute central portion of Goa and is another physiographic division with heights ranging from a hundred meters (metres) down to thirty meters (100 meters to 30 metres from the mean sea level). These formations are fairly level at the top, but in some,the places are deeply cut into gullies with sharp rivers and escarpments sloping towards the plains below them gradually or abruptly. On the coastline these plateaus and in some sort of head lands, and have scrubs and rough grass in patches on them. While the plateaus proper are flat and bare, the scarp faces have in them hollows of gullies, with vegetation of sparse stands typical of monsoon forests of thinly covered grass and shrubs, often surrounded with cashew plants. The greenery of coconut palms, paddy fields and betel gardens marks the landscape of most of the central part of Goa. The gullies in their turn present portions of evergreen vegetation with many springs and brooks of great natural beauty, feeding the rivers flowing down below them. There are clumps of grass and cashew shrubs skirted by the greenery of coconut palms and other trees and plants of a wide variety form a recurrent theme in the goan landscape.

c) The Coastal and River Plains of Goa

The state has maximum 105 kilometers of Coastline in length. The Goan Coastline Starts from the Tiracol northern tip of Pernem taluka to the Southern tip of Canacona taluka is a scenic alteration of bays, creeks, and headlands and significantly broken up by the large estuaries of the two major rivers in particular and the minor rivers (estuaries) of other rivers in general, of the bays, those of Baga and Calangute in the north and those of Colva and Betnl in the south are extensive curved stretches with almost golden and silvery white sands respectively, with coconut palm fringes all along them. The headlands are dotted with forts and landmarks for coastal and international navigation, with light houses atop them, viz., Aguada on the Northern bank of the Mondovi, with Reis Mogos in the vicinity, and Chapora fort at the mouth of the River of that name and cabo (cape) on the southern bank of the Mondovi as also touching the northern bank of the Zuari. The fortifications, now dilapidated, on the Mormagoa plateau and on the head land of Cabo de Rama in the South, are other formations on the coast of the state.

GOA – PHYSIOGRAPHY

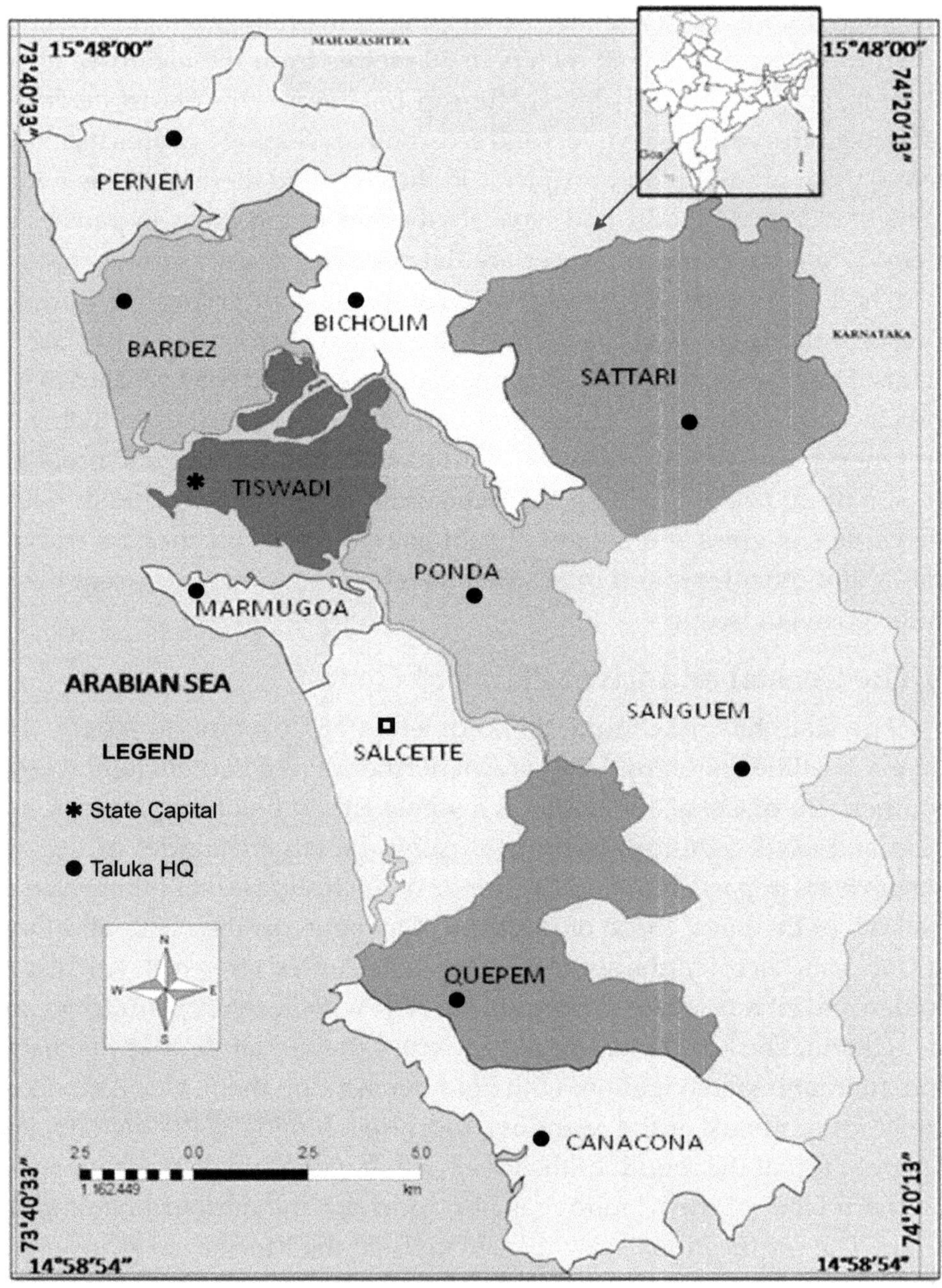

Map 2

In this interesting Mosaic of a configuration are islands of two types the rocky and the alluvial. The islands off the Mormagao Coast and Tujediva island off the coast in the south near Karwar, which are reportedly protrusions of a drowned topography separated from the mainland by a process of geological character called faulting and those in the estuary of the river Mondovi, like the Diver Chorao islands belong to the first group. The island of 2 annum or st-esterm also in the midstream of the Mondovi off Piligao and adjacent to that of Diver mentioned in the first group, may perhaps, be in the same group, as it has a rocky head of a hill, hedged in by a lot of silt that gathers in the estuary of the Mondovi. But it also keeps company with the other islands of the Kumbhayva group. These latter are exclusively created by the deposits of silt, these being known as alluvial islands on that account.

The river plains or the alluvial lowlands of Goa are the stretches of land bordering the banks of rivers, which have deposited in them the eroded materials washed down by the rivers. River Zuari and the Mondovi together constitute the most fertile area in the entire Konkan coast land. The Khazan system of alluvial formation, that is mainly centred in this area provides ideal conditions for the growth of mangroves and other floral species of botanical flowering. Where as the Chapora river in the north and the Kushvati and Sal in the south as well as of the Talpona Galgibag and Saleri rivers are known for river basins develop alluvial flats behind sand bars that form them owing to their interaction with the sea, thereby developing highly productive agricultural tracts in that portion.

GEOMORPHOLOGY

A small state of Goa has attracted the attention of several scholars in the field of geography and geomorphology.

Waggle (1993) has presented a detailed review of various attempts of explanation of geomorphic and physiographic units of Goa starting from 1911 to 1989.

Geomorphologicaly Goa can be divided into four morphological zones viz. the coastal plain with dominant marine and estuaries landforms on west, followed successively east by denudational landforms, i.e., the low dissected hills, table land, etch plain, steep escarpments and high western ghat denudational hills with deep gorges and waterfalls. The fluvial landforms are limited in their aerial extent.

Denudational hills occurring all along eastern part of Goa rise to maximum elevation of 1052m above sea level and characterized by dissected steep escarpment zone on the west, reflecting the structural trends of the folded rocks in south Goa. These hills extend westwards up to Chouri in Canacona taluka and rise to a maximum elevation of 644m. The dissected low denudational hills rise above the etch plain to a maximum elevation of 477m (east of ponda). The dissected tablelands close to the coast are characterized by laterite cover with elevation between 60-100m.

The coastal plain is characterized by short straight stretches of sand beaches with ancient beach ridges, stabilized and unstabilized coastal eolene sand dunes, small embayment (cover or pocket beaches), and cliffed headlands promontories, locally showing submerged and emergent characters. In addition, estuaries and tidal mudflats with mangrove vegetation are seen in Marmagoa and Aguada bays. These are also rocky island close to shoes; some are connected by short thin (tombolo).

DRAINAGE

The Term Drainage refers to water features that are spread unevenly in a given defined physical area. Drainage features are quite evident in the whole state of Goa with greater variations, that is, springs to large rivers. Thus, Goa being a coastal region, has plenty of water supply. There are several rivers and streams that flow through it. Most of the rivers are navigated. Surface water resources are of tremendous economic importance. There are beautiful lakes, springs water falls situated in different parts of the study area. The state has nine rivers, of which six rivers originate and flow exclusively within the state boundaries and do not have any interstate implications. However, Terekhol and Chapora rivers originate in Maharashtra while Mandove river originates on the western slopes and subsequently meanders over falls and rapids into the coastal plains (during young/initial stage), from where they tend to become sluggish (during mature stage) and then ultimately join estuary mouth, and then into the sea (last stage). Most of these rivers are subject to tidal variations and salinity up to a distance of 20-40kms upstreams from their respective mouth regions. As such these have been referred to as estuaries. Most of these rivers are excellent navigational channels and are used mostly for transporting ores through barges from ore loading jetties to Mormugao harbour for onward export to needy countries of the world. The following table shows the Rivers and their drainage basin in different parts of Goa.

Table-2 : Details of River Basins in Goa

Sl. No.	Name of the River Basin	Length within the state	Length within the salinity zone	Basin area	Average runoff
1.	Terekhol	26	26	71	164.25
2.	Chapora	32	32	255	588.35
3.	Baga	10	10	50	116.42
4.	Mandovi	52	36	1580	3580.04
5.	Zuari	145	42	973	2247.4
6.	Sal	40	14	301	694.39
7.	Saleri	11	5	149	343.04
8.	Talpona	32	7	233	515.59
9.	Galgibag	14	4	90	187.11
	Total	362	176	3702	8436.5

Source: Master plan for Madei/Mandovi river basin a report by the panel of expert (2001).

CLIMATE

Climate has been defined as aggregate of weather conditions over longer period of time on the hand, weather implies aggregate of atmospheric conditions for a moment, maximum 24 hrs.

Climatic Seasons of Goa

The state is situated within the tropic and surrounded by the Arabian sea to west and western ghats, rising to an average height of 1000 meters. The east and south region is experiencing tropical and oceanic climate conditions with profound orographic influence. Accordingly, the climate is balanced and moist throughout the year.

On the bases of various characteristics of climatic conditions prevailing all over Goa throughout the year, the climate of Goa has been divided into four main seasons.

Southwest Monsoon

South west monsoon season starts from early June to the end of September. During the season, the state is experiencing regular and sufficient amount of rainfall, reaching an average of 300 cm annually. The monsoon bursts over the state in the beginning of June and withdraws from early October. 90% of the total rainfall occurs during this season. The amount of

GOA – DRAINAGE SYSTEM

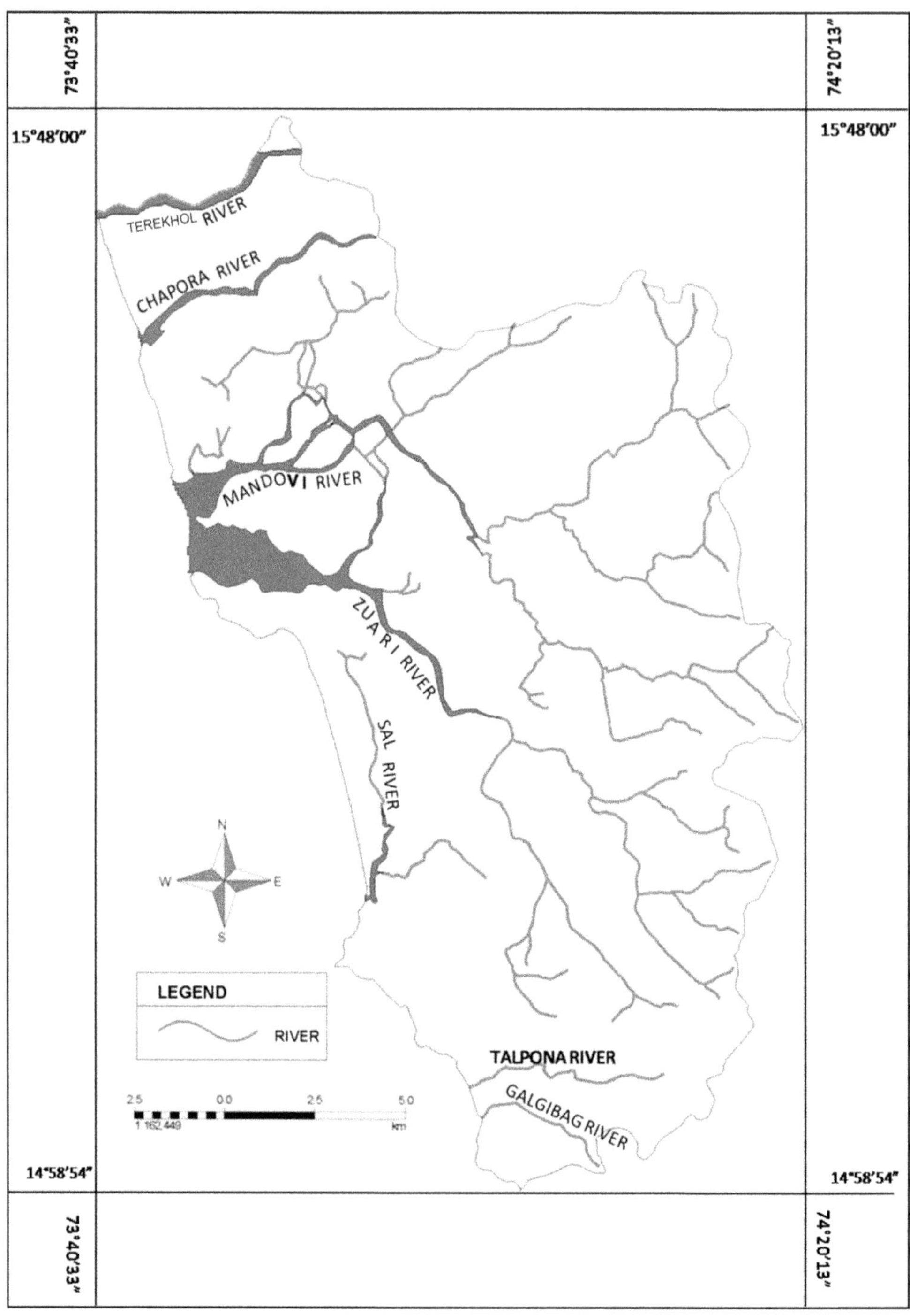

Map-3

rainfall varies from one part of the state to another. The rainfall ranges from 300 to 500 centimetres annually in different parts of Goa. The coastal belt is receiving 400 cm rainfall while the Western Ghats receives more than 500 cm rainfall annually.

Northeast Monsoon

Northeast monsoon starts from early October to the end of November. During this season northeast monsoon winds contribute remaining 10% of the total rainfall. Generally, northeast monsoon is irregular in times and places. Thus, different parts of Goa receive an occasional rainfall. During the season, the percentage of humidity is quite lower but temperatures are moderate.

Winter Season

Winter season in Goa commence from early December to the end of February. Generally, climate is good with clear skies and moderate temperature. On an average temperature ranges from 19^{0}C to 32^0 C. The winter climate is very favourable and pleasant for large number of tourists and local population of Goa.

Summer Season

The summer starts from early March to the end of May. During this season, the climate is pleasant, but discomfort may be felt in the absence of winds, particularly in the end of May, reaching maximum to 40°C, which leads to pre-monsoon showers. Similarly, the percentage of humidity varies from early March to the end of May.

Rainfall

The total rainfall received during the calendar year 2008 was 2759mm, which is much lower than that of 3,592.5 mm recorded in the previous year. Though the rainfall during 2008 was quite satisfactory, it was much below the average rainfall of 2902 recorded during the last decade.

GOA – RAINFALL DISTRIBUTION

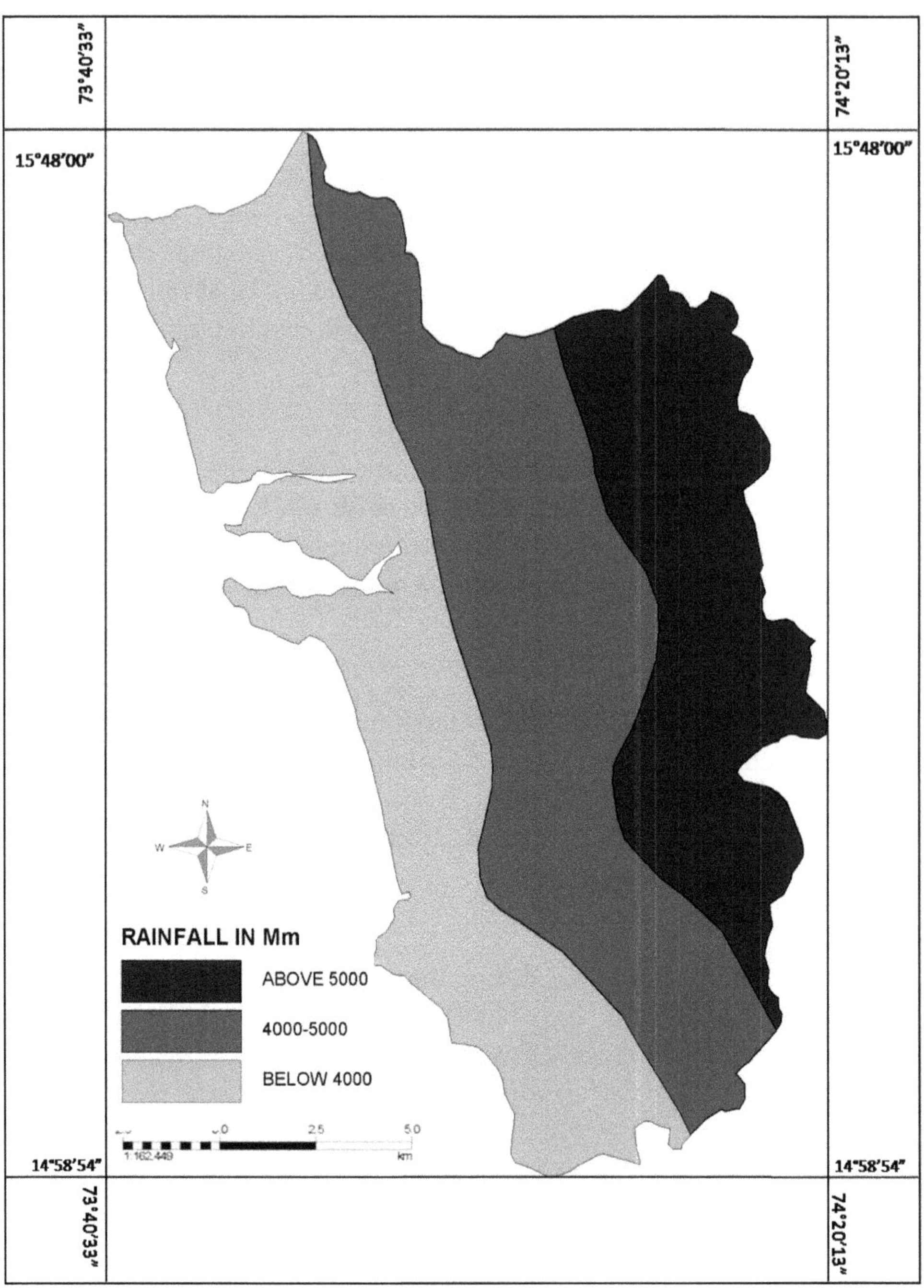

Map-2.4

Table- 3 : The Trend of Rainfall Distribution in Goa

Rainfall Recorded from 1992 to 2008	
Year	**In mm**
1992	2778.2
1993	2558.3
1994	2894.4
1995	3555.6
1996	2880.9
1997	3366.9
1998	3078.9
1999	3680.4
2000	3511.6
2001	2128.1
2002	2270.4
2003	2682.9
2004	2156.0
2005	3271.0
2006	2644.1
2007	3592.5
2008	2759.6

Source: Goa observatory. Government of Goa, 2009.

In January, February, March and December, the mean is between 2 and 3 tenths. During these months, the clouds are not cirrus. Skies with more than 8 tenths cloudy are very few.

SOIL RESOURCE

Soil can be defined as a thin layer of the earth surface which consists of unconsolidated materials, rocks, boulders, sand particles and organic matter, etc. The study area has a variety of soils in different parts. Soils can be classified as laterite (81%), alluvial and sandy. Alluvial soils are subjected to inundation by saline water and are to be protected by bunds. The coastal inland area comprises of a stretch of land with high water table, which can be exploited for irrigation and multiple cropping.

There are sandy loams to silt loams in texture, well drained and highly acidic. (5.5-6.5ph). These soils have moderate organic carbon, but are poor in phosphorus and potash. About 11% of the soils located along the sea

GOA – SOIL DISTRIBUTION

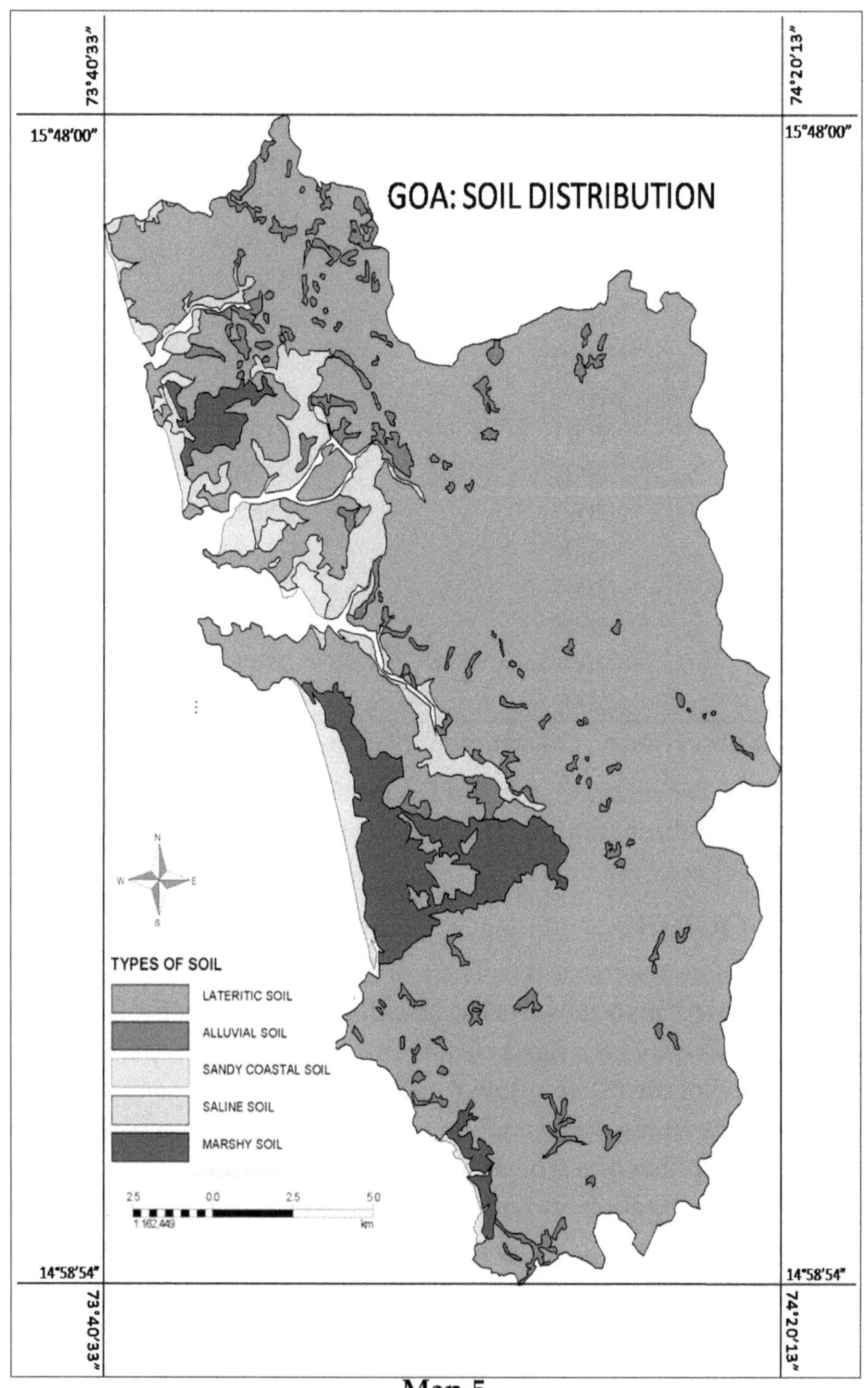

Map-5

coast and estuaries are sandy potash. They include the kher land and beach fronts. The remaining 85% of the soil are alluvial in nature. The Khazans and adjoining areas have alluvial soils. On the basis of fertility, topography of the soils, is classified into:

A. **Khazans land:** are characterized by alluvial soils, which mainly occur in low lying areas, often below the sea level, along the estuaries. This land is used for monsoon paddy crops, followed by Rabi vegetables.

B. **Kher land:** This land is at low elevation above the sea and is having a high water table. Aerable sand loams are for multiple cropping through irrigation.

C. **Morod land:** Is largely characterized by laterite soils confined to upland or terraced fields suitable for horticulture crops or as well as agricultural crops.

The soils found in the state are well drained, but have contents of nitrogen and phosphorous. Although dense vegetation and grass have contributed to high contents of organic matter (0.5 to 1.5% organic carbon) in several soils. The pH of the soils is 4.5 to 6.5. Soils in general are productive and respond well to irrigation and fertilizer management.

FOREST

Forest is a Natural resource provided by Nature as a free gift to man on the earth surface. In other words, forest is a plant society, which consists of large and small varieties of trees, grasses, shrubs, bushes and creepers. The Term forest has been derived from the Latin word "Foris", meaning outside, a village border fence.

Table- 4 : Goa – Area under Forest

Geographical area	: 3702.00
Forest area (sq.kms.)	: 1424.46
Government forest (sq.kms.)	: 1224.46
Private forest (sq.kms.)	: 200.00 approx

The study area has a wide range of forests, covering 1424 sq.km area. Out of the total geographical area of 3702 sq.km, about 1424.46 sq.km constitutes the forest. Of this, 1224.46 sq.km has been classified as government forest, and the rest private forest.

Table-5 : District wise Distribution of Forests

Name of District	Area (sq.km.)
North Goa	354.48
South Goa	869.98

The following table shows the distribution of forests in Goa.

Table-6 : Taluka wise Distribution of Forest in Goa

Goa District (Taluka)	Geographical area in HA	Forest area in HA
Tiswadi	16,612	178.00
Salcette	27,719	8.00
Bardez	26,780	-
Mormagoa	7,831	-
Ponda	25,228	5012.46
Bicholim	23,633	808.23
Pernem	24,200	1343.26
Quepem	43,731	11490.73
Sanguem	88,660	56924.83
Canacona	34,736	18581.79
Satari	51,284	28099.15

The above table depicts distribution of forests in various talukas of Goa. Sanguem taluka has recorded largest area covered by thick forests of evergreen and semi-evergreen about 56.924.83 hectares followed by Ponda 5012.46 hectares. The sparse and dense forest has been noticed in Quepem taluka with just 8.00 hectares due to excessive urbanization. The talukas of Canacona, Quepem, Tiswadi, Satvi, have moderate concentration of forest of mixed variety. Pernem taluka also falls in the zone of less concentration of forests. Hence, the forest have unevenly distributed and have rich Bio-diversity.

The following table shows administrative units in the forest department.

Table-7 : Administrative Units in the Forest Department

No. of Territorial Divisions	2
No. of Wild life divisions	1
Total No. of Divisions	9
No. of Territorial Ranges	11

Table-8 : Forest Area by Legal Status

Reserve forest (sq.km.) u/s 20 of Indian forest Act 1937	236.82 sq.km
Proposed reserved forest (sq.km.) under section 4 of Indian Forest Act 1937	722.63
Unclassed forest (sq.km.) declared under section 18 of Wildlife (Protection) Act	137.11
Unclassed forest (sq.km.) identified for section 4 notification	96.21
Balanced Unclassed forest area (sq.km.)	31.69
Total	**1224.46**

Source: Department of Forest and Environment, Government of Goa, (2009)

Forest Types Found in Goa

The vegetation of Goa is typical of the Western Ghats (Southern Maharashtra and Karnataka). The narrow coastal plains lead eastward to hills ascending about 1200 m with altitudinal zonation. The vegetation shows a spectrum of variability from west to east, the gradient being correlated with ascending contours from sea coast to the crest line of the Ghats.

Vegetation of Goa can broadly be classified into:

i. Estuarine vegetation consisting of mangrove species; along the narrow muddy banks of rivers,

ii. Stand vegetation along the few coastal belts,

iii. Plateau vegetation confined especially to the low altitude of the Ghats,

iv. Semi-evergreen and evergreen forest limited of patches along high altitude of the Ghats. Their details are as follows :

- Estuarine Vegetation of Mangroves along Swampy River Banks:

In Goa, this vegetation is distributed in a 3 sq. km. area. Botanically this zone is characterised by peculiar root formation (stilt roots of Rhizophore, Pneumatophores in Avicenna, Knee root in Bruguiera etc). These occur mostly in sheltered bays, and are found in areas which are covered by salt or brackish water at high tidal streams and borders of lagoons and estuaries more or less protected against heavy wave action and winds. The main mangrove localities in Goa are Maxem in Canacona, Durbat, Panaji and Akshi, Cortalim.

- Strand and Creek Vegetation along Coastal Belt.

Most of the coastal regions of Goa are rocky with projecting ridges as well as rocky boulders and, consequently, the strand vegetation is limited to a few patches of narrow strip, bordering the Arabian Sea. The vegetation along the south bank of river Mandovi near Panaji belongs to this category.

Trees species are: *Pongamia pinnata, Thespesia pupulnea, Calopyhllum iniphyllum, Cerbera manghus,* and *Pandanus tectorius*. The other associates such as shrubs are *Derris trijoliata* and Caesalpinia; crista intermixed with herbs *Sesuvium portulacastrum,* phyla nudiflora, *Arthrocnemum indicum, Melanthera biflora,* sedges like *Cyperus arenarius* and *Fimbristhis schoenoides* and grass *Spinifex littorens.*

- Plateau Vegetation along Undulating Terrain and Hills:

A major portion of Goa belongs to this category extending from 50-200 mt and further divided into two types viz. a) open scrub jungle b) moist deciduous forest

♦ Open Scrub Jungle:

This type of vegetation occurs from Panaji to Cortalim, from Cortalim to Margao, and from Bicholim to Sanquelim. *Anacardium occidentale* is cultivated on an extensive scale. Several eroded waste lands sustain patchy vegetation composed of dry deciduous elements such as Carissa, *Calycoopteris floribunda,* Wood *Fordia fruiticosa, Grewia tilfolia, Vitex negundo* and species of *Calotropis, Zizyphus, Cassia, Ixora, Acacia, Albizia, Terminalia* and *Crotalaria.*

♦ Moist Deciduous Forest:

This is the main forest type of Goa, covering more than half of the catchment. This type of forests occurs around Tudal, Orolofond, Butpal, Molem, Codal, and Ambiche Gol near Valpoi, Anmode Ghat and Canacona. Predominant species are *Terminalia crenulata, T. bellarica, T. paniculate, Hagerstroemia lanecolata, Adina cordifolia, Albizia lebbeck, A. procera, Mitragyna parvifolia Holoptella integrifolia, Trewia nudiflora, Dillena pintagyna, Semicarpus anacardium, Mallotus philippensia* and *Stereospermum colais.*

◆ **Secondary Moist Mixed Deciduous Forest :**

This type has formed due to moist conditions, resulting in secondary origin of past shifting cultivation. Trees found in this type are knotty and of coppice origin. The main species found are *Terminalia crenulate, T. chebula, Adina cordifolia, Alstomia scholaris, Hannea coromandelica, Bombax Ceiba, Careya arborea* and *Dillenia pentagyna.* Common associates are *Xeromphis spinosa, Ziziphus xylopyrus,* z. caracutta, *Calycopoteris floribunda, Helicteres isora* and *Moullava speicata.*

◆ **Sub Tropical Hill Forest:**

These forests have formed due to past "kumeri" cultivation. *Syzygium cuminni* and *Cinnamomum verum* are common occurrence. *Caryota urens* is the most common palm over such type. In the second story Carvia Callosa (*Stribukabtge calllousus*), *Elaeagnus coonferta* and *Capparis* spp. are found, sendoxytenanthera is of frequent occurance.

- Semi Evergreen and Evergreen Vegetation Along Upper Ghats:

◆ **Semi Evergreen Forests:**

This type intermingles between the tropical evergreen and the moist deciduous forest mostly above 500m, bordering the contiguous forest of Ratnagiri district, or the North Kanara district in the south. This forest occurs at Amboche gol, Molem, Butpal and Nadquem. Species composition is of *Artocarpus hirsutus, A. gemeziamus, Calophyllym* spp. *Sterculia guttata, Kyida calycina,* Hagerstroemia, Microcarpa, *Pterospermum diversifolium, Garcinia indica, Diospyros montana* and *Macranga pettala.*

◆ **Lateritic Semi Evergreen Forests:**

The soils are typically lateritic, shallow dry and open. *Xylia xylocarpa* is the prominent tree species of this type with other associates like *Pterocarpus marsupium, Grewia tillifolia, terminalia Paniculata, schleichera, Oleosa, Careya Arborea* and *Bridelia retusa* and *Strychnos nux- vomica.* The ground flora is typically represented by *Calycopteris floribunda* and *Holarrhena pubiscens.*

◆ **Evergreen Forest:**

In deep gorges and depressions, along the nallahs and streams with congenial soil and moisture conditions, the evergreen species occur. The evergreen species occur with a composition of *Calophyllum, inophyllum, Garcinia indica, Canarium strictum, laphospetalum wightiamum, myristica suppknema altenuata, Chroisophyllum acuminata, Pakaqyuyn ekkuotucum, Artocarpus gomezianus, Mangigera indica, persea macrantha, Mimusops elengi, Hopea Wightianum, Olea diocia, Hydnocarpus laurifolia, Syzyagium cummi, phillippensis, Ficus* spp.

Chapter-2

Demography – Population Structure of Goa

The term population has been derived from the Latin word "populus", meaning group of people living in a given defined physical area. Population, surely, is an important aspect of any state or country in the world. Manpower or human resources are both instrument and goal of economic development. As an instrument, human resources supply an essential service with the combination of other services. On the other hand, all development activities in the economy is undertaken to provide progressively better standard of living to human beings. Thus, human resources as units of production, consumption and accumulation, account for the total output of the economy. The size of population, therefore, is a crucial determinant of economic development of Goa state.

The survey conducted by the Directorate of census operation, Goa puts Goa's population at 13,47,668 inhabitants as against a figure of 11,69,793 inhabitants as per the 1991 census. The most promising aspect of the census is that Goa's decadal growth of 34.77% during the period 1961-71, tapered off to 26.74% during 1971-81 and further to 15.96% during the period 1981-91. The growth of 34.77% during the first decade after Goa's liberation, compared to the national growth rate of 24.78%, had caused apprehensions in Goa in respect of the heavy influx of outsiders into Goa. The 1991 figures have come as a big relief to those who had nourished fears that the heavy influx of outsiders experienced during 1961-71, if maintained, would bring Goans into a minority in their own land. It appears that such an influx has tapered off to a sizable extent, and Goa's commendable achievement should be largely due to small families and late marriage norms adopted by Goans. Improvement in the literacy percentages, especially among families has contributed substantially to restricting population growth in Goa.

Another significant achievement of Goa, as revealed by the 2001 census results, is the high literacy percentage of 82.0%, which is a big jump from 35.41% prevailing in 1960 just before liberation and which is much higher than the national average of 64.8%.

The four coastal talukas of Tiswadi, Mormugao, Salcete and Bardez, which accounts for about one fifths of Goa's total geographical area sustain 57.79% of the state population whereas, Sanguem taluka with an area of 23.25%, accounts for only 5% of the population Urban population, as against 32.46% and 26.44% in 1981 and 1971 respectively, the share was only 14.8% in 1960.

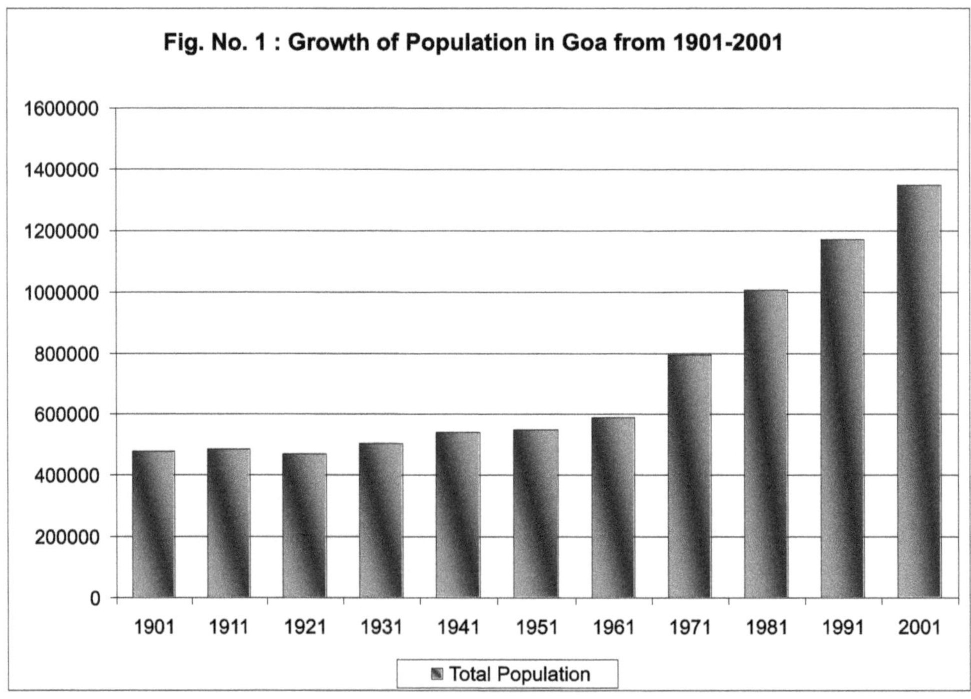

The following table shows various trends of demography in Goa during the period 1901-2001.

Table-9 : Trends of Population

Years	Person	Decade	% of Decade	Males	Females
1900	475513	-	-	227393	248120
1910	486752	+11239	+2.36	230923	255829
1921	469494	-17258	-3.55	221429	248065
1931	505281	+35644	+7.62	241936	263345
1940	540925	+35644	+7.05	259591	281334
1950	547448	+6523	+1.21	257267	290181
1960	589997	+42549	+7.77	285625	304372
1971	795120	+205123	+34.7	401362	393758
1981	1003141	+208021	+26.74	510152	497597
1991	1169793	+166652	+16.08	594790	575003
2001	1347668	+177875	+15.21	687245	660420

Source: Directorate of Census Operation Goa, 2001.

DEMOGRAPHIC STRUCTURE OF GOA

The study of population structure of Goa mainly consists of the following:

a) Growth of population,
b) Density of population,
c) Sex ratio.

The above mentioned concepts have been studied over a period of 100 years from 1901 to 2001. Their details are as follows.

Table-10 : Growth of Population in Goa

Year	Total Population	Decade Variation	Percentage Decade Variation
1901	4,75,513	-	-
1911	4,86,752	+11239	+2.36
1921	4,69,494	-17,258	-3.55
1931	5,05,281	+35,787	+7.62
1941	5,40,925	+35,644	+7.05
1951	5,47,448	+6,523	+1.21
1961	5,89,997	+42,549	+7.77
1971	7,95,120	+205123	+34.77
1981	10,07,749	+2,12,629	+26.74
1991	11,69,793	+1,62,044	+16.08
2001	13,47,668	+177875	+15.21

Source: Directorate of Census Operation Goa, 2001.

a) Growth of Population in Goa

Growth of population is an important component of population studies. It refers to explosion of population due to various factors, i.e., high birth rate and low death rate. It is more highly concentrated in urban centres than in rural areas. Maximum density can be seen due to overcrowding which is taking place day by day in major towns and cities of Goa.

According to 2001 census report, total population of Goa is about 13,47,668, out of which 6,87,248 are males and 6,60,420 females. It accounts for 80% of the total population of the state. The table 2.10 and simple line graph shows the growth of population in Goa during various decades from 1901 to 2001. From the table and simple line graph it appears that Goa had just 4,75,513 population in 1901, which is less than 60% compared to the present size of population in the year 1991, out of which 2,27,393 were males and 2,48,120 females.

According to the law of nature, everything changes from time to time and place to place. And so the structure of population of Goa started taking different shapes gradually, and it experienced growth of population of +2.36% by the end of first decade in the year 1901, as a result of which the size of population increased to 4,86,752. Goa has experienced sudden decline in growth of population which was about -3.55% during the period 1911 to 1931. The state experienced tremendous growth by +7.62%, which is more than 3 times than earlier due to sudden increase in birth rate and control over death rate by improvement in medical facilities and transportation, which are mainly responsible for the growth of population in the state and the figure reached to 5,05,281.

During the period from 1931 to 1941, there was an increase of +7.05% growth of population, which showed a significant growth of population, but during the period between 1941 and 1951 there was a sudden decline in growth by +1.21%, a small increase in the growth of population due to spread of educational facilities in most of the parts of Goa. Literacy campaign had been launched in rural and urban areas of Goa to create awareness among the people.

The real growth of population had taken place in the state from 1951 to 1991. The size of population, which was 5,47,448 in 1951 rose to 11,69,793 by 1991 to 13,47,668 in the year 2001. A net increase of 12.0% growth was experienced by the state. Especially during the period from 1961 to 1971 there was high growth rate even than before it was about +34.77% a tremendous increase among several decades experienced by the state.

This high rate of growth was the result of migration that took place from neighbouring states like Karnataka, Maharashtra and also from the different parts of the country.

After the liberation of Goa from Portuguese on 19th December 1961, a large scale movement of people had taken place to avail themselves of the opportunities available in the various fields of the state to support their life. Real expansion of mining activities attracted, to some extent, labour from neighbouring provinces.

The process of growth of population continued after the post-liberation period due to the availability of plenty of resources in the state. Gradual developments have taken place in the economy of Goa by achieving considerable growth in the various sectors, which was a sort of an encouragement to the local people and outsiders to participate in the economy of the state.

The following table shows population of Goa in the district of North Goa, South Goa and decadal growth during 1900-2001.

Table-11 : District wise Population of Goa and Decadal Growth Rate

Sl. No.	Year	North Goa District	South Goa District	Goa Total	Decadal Growth (%)
1	1901	294074	181439	475513	-
2	1911	306323	180429	486752	+2.36
3	1921	255039	181455	469494	-3.55
4	1931	313614	191667	505281	+7.62
5	1941	336628	204291	540925	+7.05
6	1951	330874	216574	547448	+1.21
7	1961	349667	240330	589997	+7.77
8	1971	458312	336808	795120	+34.77
9	1981	568021	439728	1007749	+26.74
10	1991	664804	504989	1169793	+16.08
11	2001	758573	589095	1347668	+15.21

Source: Directorate of Census Operation (2001).

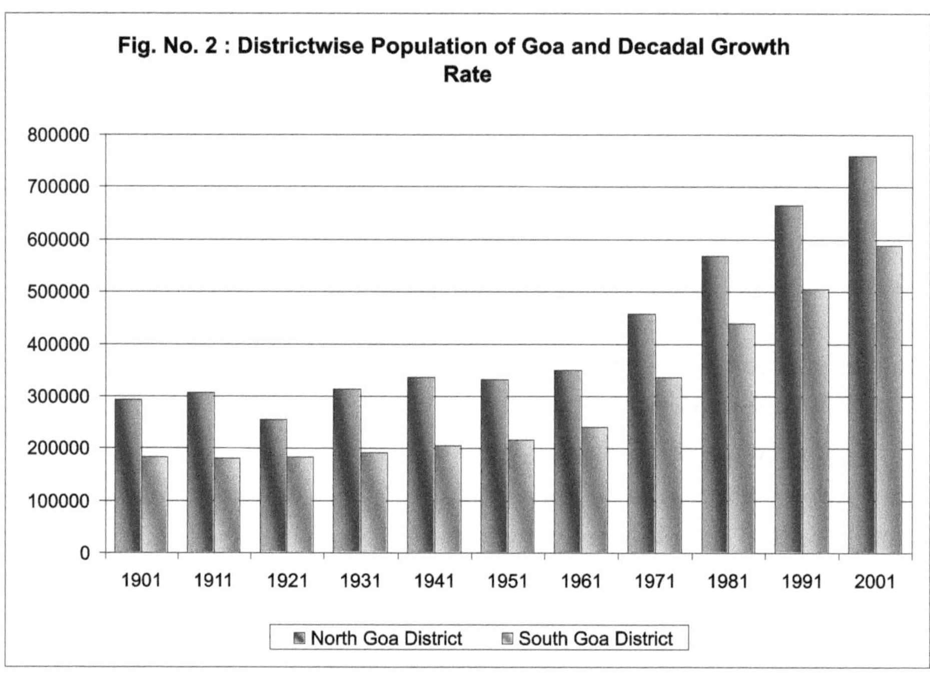

Out of the total population of 13,47,668, north Goa district has 56.29% and south Goa district has 43.71%. The decadal growth rate in population has declined from 16.08% during 1981-1991 to 15.21% during 1991-2001.

b) Density of Population

Density of population refers to number of persons living in per sq.kms area or per unit land area in the state. Density varies from one place to another, from one region to another. There are various factors influencing density of population such as physical, economic, social and religious, etc.

Table 10 and simple bar graph shows the trend in density of population of the state during various decades. The density of population of state was very low as 128 people living in per sq.km area in the year 1901, which rose to 364 people per sq.km in 2001. This represents 290% of increase in density of population in 1991. Population in the state is not evenly spread, and variation in the density within all talukas of the state is very wide. According to the 1991 census report, it is as low as 68 persons living in per sq.km area in Sanguem and as high as 1104 person living in per sq.km in Mormugao taluka. On the basis of different density, all talukas have been classified into different density zones whereas Pernem, Bicholim, Sattari, Quepem,

GOA – DENSITY OF POPULATION (2001)

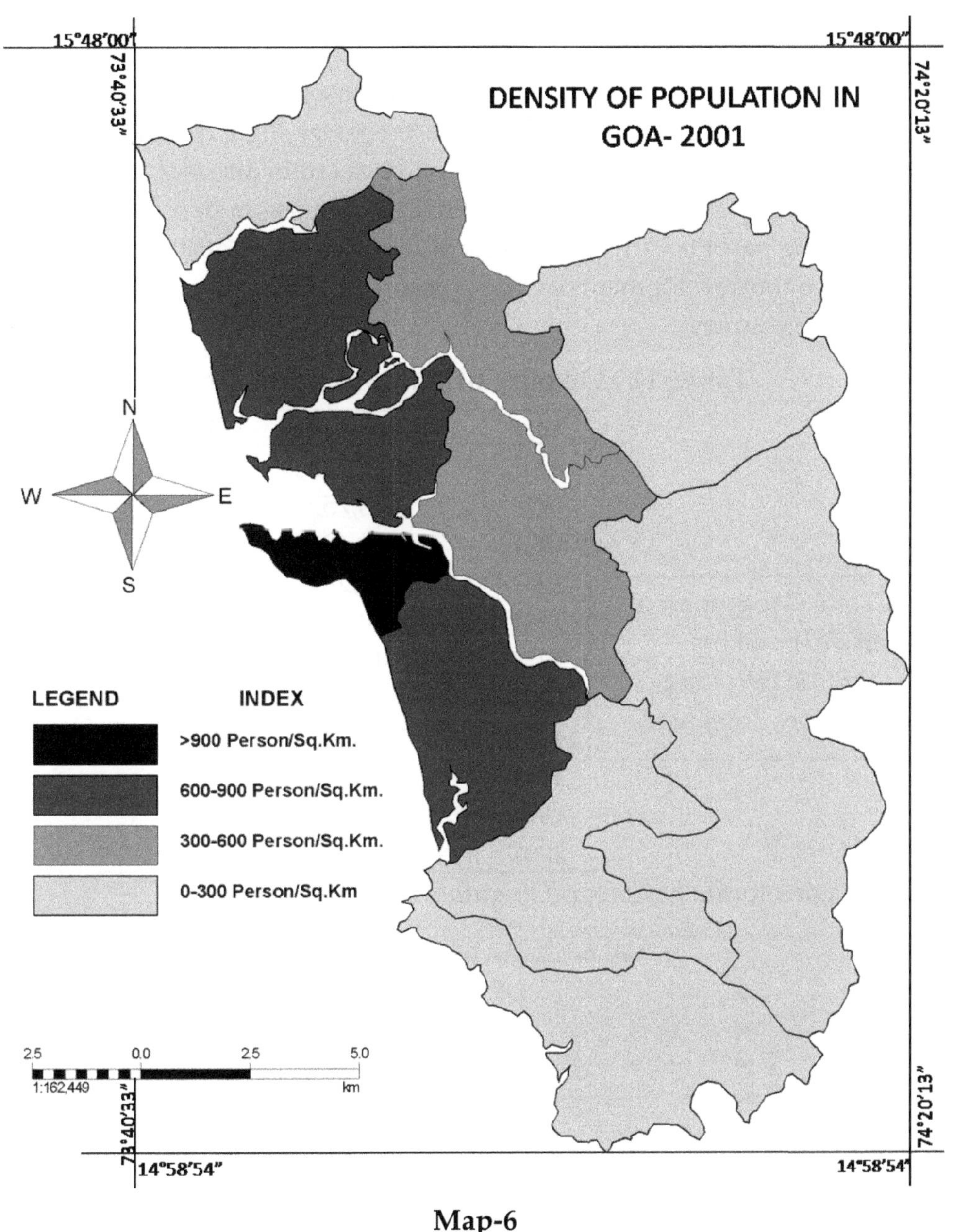

Map-6

Canacona comes under medium density zone wherein more than 100 and less than 300 persons per sq.km area can be found and Bardez, Tiswadi, Ponda, Mormugao and Salcete belongs to high density zone wherein more than 300 person are living per sq.km area.

The density of population in the state increases rapidly over the several decades. But during the period 1911-1921, there was slight decline in density of population due to high death rate caused by certain diseases, i.e., cholera, malaria, small pox, etc. Further tremendous increase in density has taken place during past liberation period from 1961-71 almost 74% increase in the state due to the development of various sectors and migration from different parts of the country.

Table-12 : Density of Population in Goa

Year	Total pop.	Density /sq.kms
1901	4,75,513	128
1911	4,86,752	132
1921	4,69,494	127
1931	5,05,281	137
1941	5,40,925	146
1951	5,47,448	148
1961	5,89,997	159
1971	7,95,120	215
1981	10,07,749	272
1991	11,69,793	316
2001	13,47,668	364

Ds Total Geographical-Area 3701 sq.kms
D DENSITY= total-population/total area

Source: Directorate of Census Operation (2001)

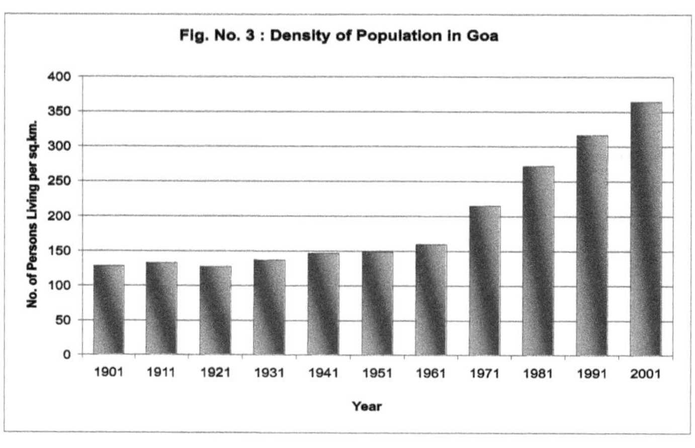

SEX RATIO

Sex ratio refers to the proportion of female's population to the male population. It is described in terms of 1000 persons i.e. number of females per 1000 male's population. This can be obtained by using following equation:

Sex ratio = female/males*1000

According to various census reports, sex ratio is uneven in all talukas of Goa due to biological and other factors. It differs from one taluka to another. During 1901 to 1961, the sex ratio was very high in the state, i.e., 1220 females per 1000 male population in the year 1921. But after 1961 onwards, during post-liberation period, the ratio of females had declined due to various factors such as improvement in medicine facilities, and literacy campaign launched in rural and remote areas of the state. Increase in the rate of literacy, living standard of the people and improvement in scientific techniques in identifying sex at private hospital lead to decline in sex ratio over the past three decades. Finally the state has low sex ratio about 961 per 1000 by 2001, according to census department, Goa.

Table-13 : Sex Ratio

Year	Total Population	Males	Females	Sex ratio=f/m*100
1901	4,75,513	2,27,393	2,48,120	1091.15
1911	4,86,752	2,30,923	5,55,829	1107.85
1921	4,69,494	2,21,429	2,48,065	1220.29
1931	5,05,281	2,41,936	2,63,345	1088.49
1941	5,40,925	2,59,591	2,81,224	1083.75
1951	5,47,448	2,57,267	2,90,181	1127.93
1961	5,89,997	2,85,625	3,04,372	1065.63
1971	7,95,120	4,01,362	3,93,758	981.05
1981	10,07,749	5,10,152	4,97,597	975.38
1991	11,69,793	5,94,790	5,75,003	966.73
2001	13,47,668	68,7248	6,60,420	961

Source: Directorate of Census Operation (2001).

GOA – TALUKAWISE SEX RATIO (2001)

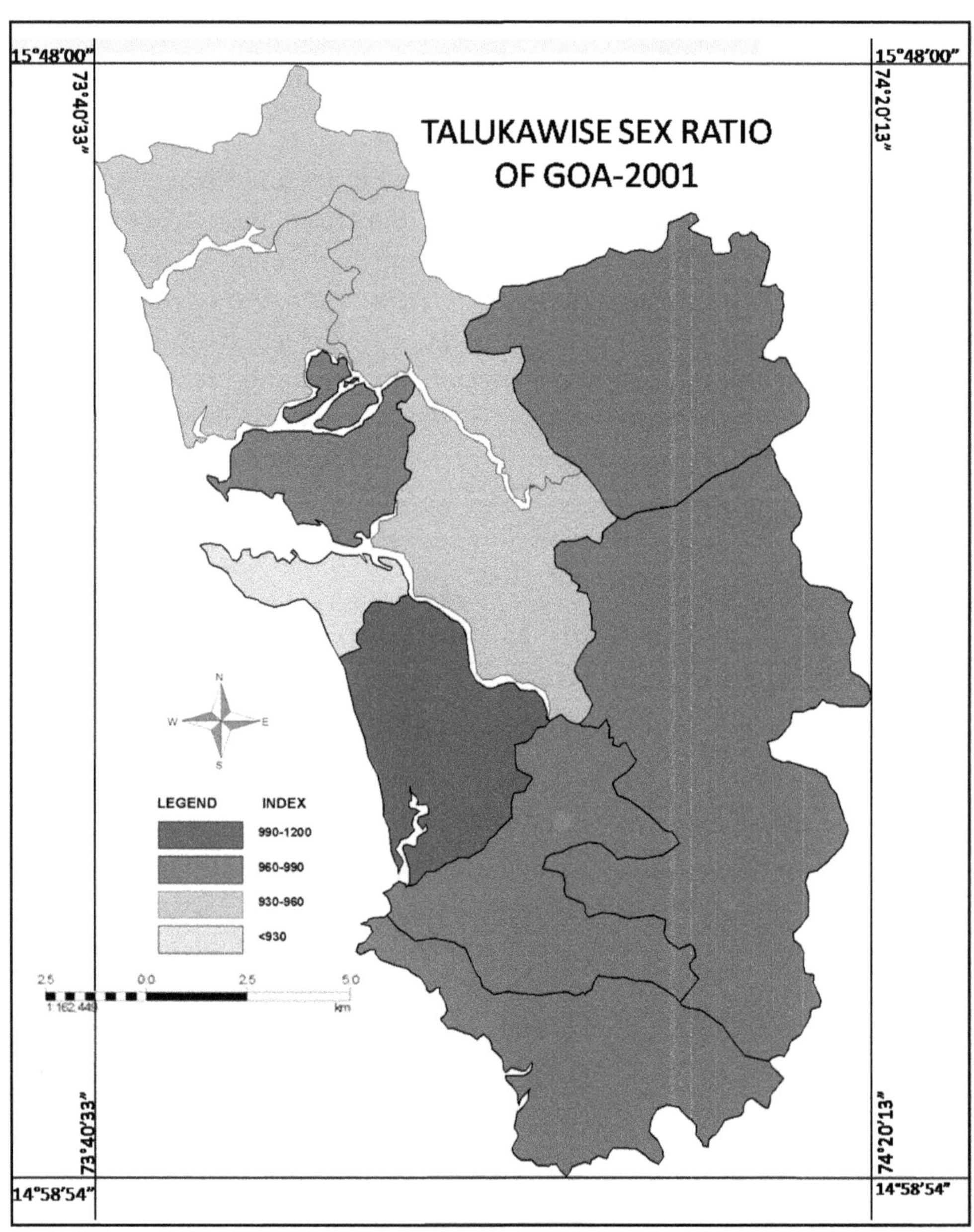

Map-7

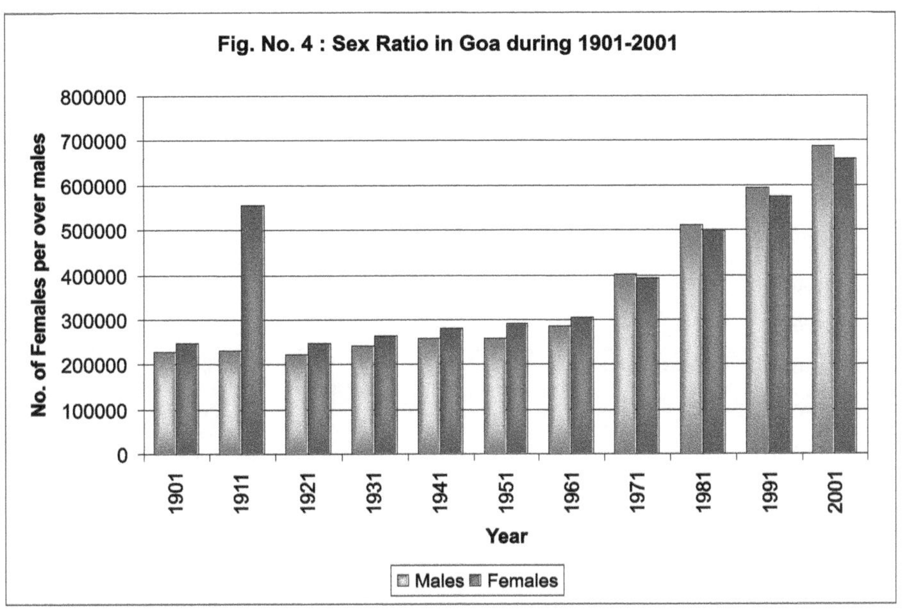

OCCUPATIONAL STRUCTURE

It refers to proportion of people engaged in different economic activities in the state. A major problem being faced is the exodus of rural population to urban areas for employment, especially for white collar jobs. As per 2001 population census, 49.77% of Goa's population is in urban areas. It is in this context that the government of Goa proposes to ensure the sustainability of the village economy by considering the village as the focal point for promoting rural economic activities, like agriculture, horticulture, floriculture, dairy development, fisheries, etc., so that people living in the village can earn their livelihood in the village itself, and do not feel that they have to migrate to urban areas in search of employment.

Table-14 : Change in Occupational Structure

Sl. No.	Category	1991	2001	Decadal growth %
1.	Cultivators	68636	50395	-26.6
2.	Agricultural laborers	44775	35806	-20.0
3.	Workers in household industry	9835	14746	+49.9
4.	Other workers	289490	421908	+45.7
5.	Total workers	412736	522855	+26.7
6.	Non-workers	757057	824813	+8.9

Source: Economic Survey (2008-09).

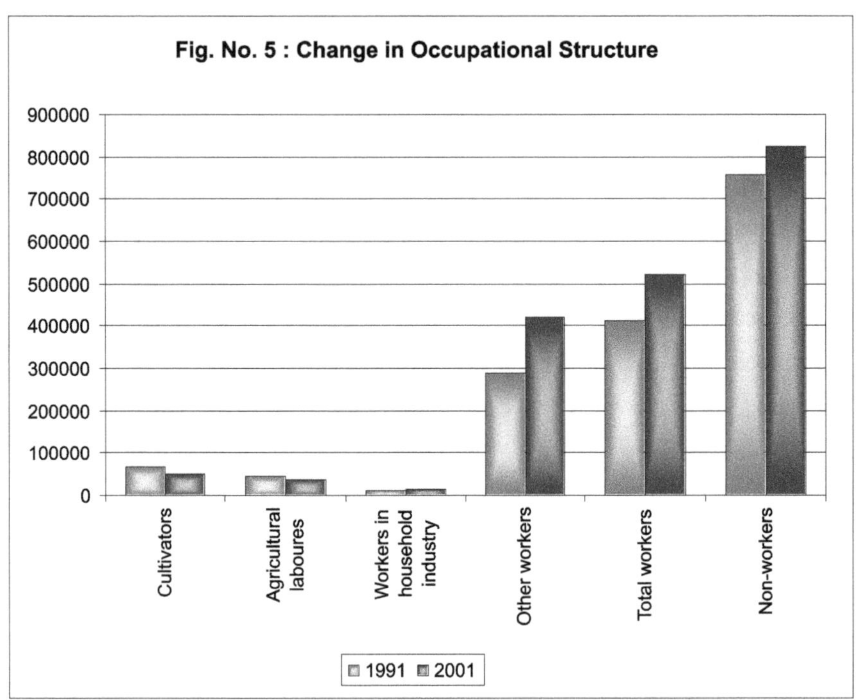

Statistic reveals that agriculture and horticulture products, like rice, sugarcane, vegetables, fruits, flowers, etc., are being imported from other states in huge quantity. The state has tremendous demand for agricultural products. There is a need to improve the economic viability of this sector. The importance of agriculture as a source of gainful employment is declining over the years on account of various factors like un-remunerative prices and lack of proper support systems for storage and marketing.

In order to address this problem, the govt. has set up "krishi ghars", which will act as collection, sorting, storage and selling centres for agriculture and horticulture products at various locations covering most talukas in the state.

Table-15 : Urbanization Trend in Goa 1950-2001

Census years	Urban population to Decadal growth of urban total Population	Population (Per cent)
1950	12.96	-
1960	14.80	23.12
1971	25.56	132.73
1981	32.03	58.82
1991	41.01	48.63
2001	49.77	39.42

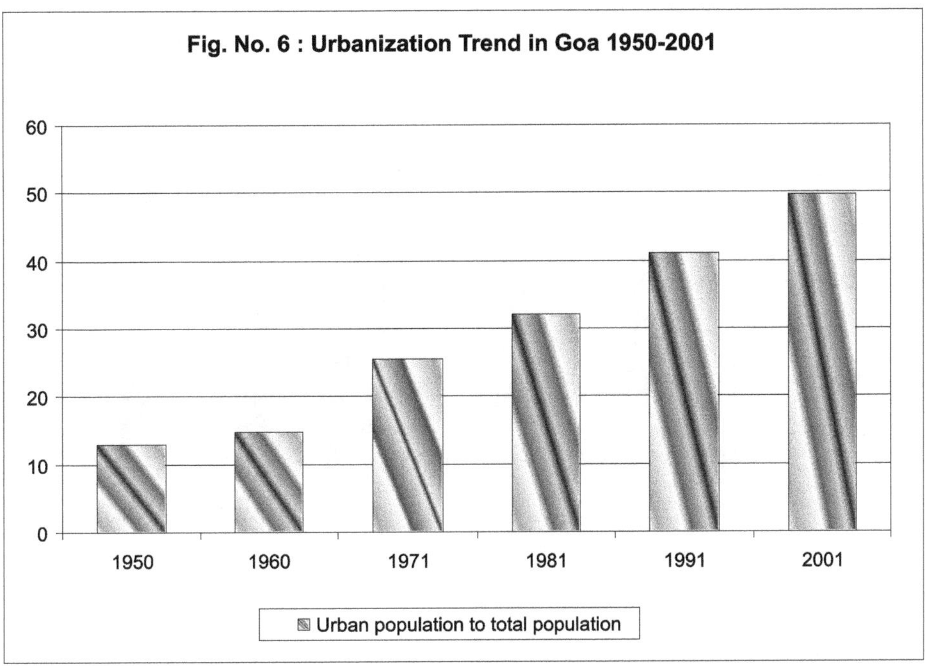

Post liberation period witnessed a heavy urbanization in Goa. As per the 2001 census, nearly half the state is urbanized. From the above table, it is evident that there is a steady increase in urban population after liberation.

Table-16 : Rural Urban Composition of Population in 1950-2001

Census year	Rural popl.	Urban popl.	Total Urban population as percentage to total population
1950	478891	68557	547448 12.96
1960	502668	87329	589997 14.80
1971	591877	202343	795120 25.56
1981	690041	479752	100774932.03
1991	690041	479752	116979341.01
2001	677091	670577	134766849.76

Source: Economic Survey, 2005-06.

As per the 2001 census, 49.76% population of Goa lives in urban areas. The post-liberation period witnessed heavy urbanization in Goa.

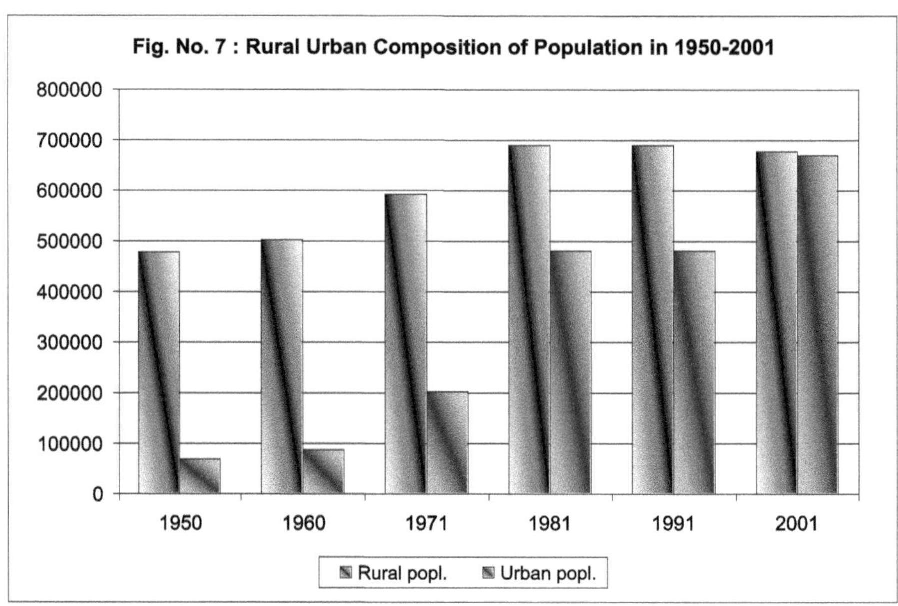

LITERACY

The literacy rate in Goa is 82.01%. The female's literacy is 75.37% and male's literacy 88.42% as per the 2001 census. Literacy rate in percentage is given below:

Table-17 : Literacy Rate in Percentage

Year	Male	Female	Total
1960	48.70	22.80	31.23
1971	54.65	35.79	45.31
1981	65.99	48.29	57.25
1991	83.64	67.09	75.51
2001	88.42	75.37	82.01

Source: Economic Survey, 2005-06.

Goa has one of the highest literacy rates among the states in India. As per 2001 census, Goa is ranked fourth next to Lakshadweep, Mizoram and Kerala. However, there continues to exist a gap of 13% males and females literacy rates.

Chapter-3

Settlement Characteristics of Goa

INTRODUCTION

Settlement geography is a study of the form of the cultural landscape. It is a science of systematic inquiry of occupancy features distributed over space with differentiation in relation to man.

Settlement is the topographic expression of the grouping and arrangement of two fundamental elements, i.e., houses and highways branches.

Shelter is one of the most important basic necessities of human being. Even the naked saints or pygmies need sound sleep, which is the psychological necessity of living beings at some place. Man also needs some sort of shelter for safe rest, i.e., tree branches, caves, pits or rock cut hiding places.

These shelter places become the most concrete expressions of human cultural activity, and assume various forms as well as names. House dwellings, group of dwelling houses, abodes and habitations, all forms of human habitat more specifically constitute settlement.

The term settlement denotes "Territoriality". Thus settlement refers to the grouping of people and houses into hamlets, villages, towns and cities (Dickenson Pitts).

Table-18 : Goa – Settlements Profile (1971-2001)

Population Size (Nos)	No. of Settlements				Percentage to Total Inhibited Settlements			
	1971	1981	1991	2001	1971	1981	1991	2001
50,000-99,999	1	3	3	3	0.25	0.75	0.78	0.79
20,000-49,999	3	1	1	2	0.76	0.25	0.26	0.52
10,000-19,999	2	4	9	18	0.51	1.00	2.33	4.74
5,000-9,999	35	35	41	38	6.35	8.73	10.62	10.02
2000-4,999	76	88	95	88	19.29	21.94	24.61	23.21
1000-1999	80	96	77	75	20.30	23.94	19.95	19.78
1-999	207	174	160	155	52.54	43.39	41.45	40.89
Total Inhabited	394	401	386	379	100.00	100.00	100.00	100.00
Total Uninhibited	8	9	15	12				

Source: Department of Census (2001), Goa. Prabir Kumar Rath (1998), Development of Goa, Ph.D. Thesis.

In other words, settlement is the place where one person or more dwells regularly or the act of establishing a permanent resident (K.H. Stone).

The study area has significant settlement characteristics as it has 359 villages, out of which 347 villages are inhabited and 12 are uninhabited. There are 14 statutory towns and 30 census towns located in different parts (2001 census report).

Goa has been divided into two main districts, i.e., North Goa and South Goa. The North Goa district comprising six taluks, 213 villages out of which 209 villages are inhabited and 4 are uninhabited. There are 7 statutory towns and 20 census towns. The South Goa district comprises 5 taluks, 146 villages out of which 138 are inhibited villages, 8 are uninhibited villages, 7 statutory towns and 10 census towns.

There are 2,94,812 normal occupied residential houses in the whole study region, followed by 1054 institutional and 1393 houseless settlements.

The following table shows the trend of structure of settlements based on the population size during the period 1971-2001 in Goa.

From the table 18, it has been observed that the settlements of the state have been classified in descending order of their size and rank. The settlement size distribution has been obtained by population of settlements during various decades, and a differentiation was made between rural and urban settlement. Vasco, Margao and capital city Panaji are highly urbanized with a population size of more than 50000 persons. Hence, these are developed urban settlements, whereas 155 villages having less than 1000 population which are called rural settlements (2001 census report), i.e. Surta. There are two settlements approaching towards large scale urbanization. One is Mapusa and another is Ponda, wherein population is growing rapidly over the years. Their size could be more than 30000 inhabitants, which has lead to tremendous infrastructure and housing developments in the study area.

Chapter-4

Economic Characteristics of Goa

AGRICULTURE AND HORTICULTURE OF GOA

The term agriculture has been derived from Latin word "Ager" means soil and "Cultura" means cultivation. Thus, tilling of the fields or cultivation of soil is called agriculture. The soil has been cultivated by man over the longest period of time on the earth surface in order to obtain variety of food materials, raw materials and fruit and vegetables needed by man for consumption purpose. In other words raising the plant life from the soil is called agriculture. Development of Agriculture ensures greater socio-economic development in the state/country.

Before liberation, Goa economy was mainly based on mining industry, yet over the last 4 decades assessing the transformation take place, it would appear to any one that agriculture in Goa is in a state of decline. Goa does possess an appropriate natural environment for agriculture. The soil and abundant monsoon are what any agriculturist craves for. There are green fields on both the sides of our roads, vast coconut plantation, hillocks covered by cashew shrubs, large spices farms that have turned into tourist attraction.

The state is witnessing a decline of growth in agriculture sector due to the infrastructural development, industrialization, mining and other services. The decadal census results indicate declining preference for agriculture in the state. In the 1961 census, more than 60% of the work force was engaged in agriculture. However, successive decadal census results of the state depict a declining trend in work force pursuing agriculture.

The following table shows the trend of decline in agriculture preference by the people of Goa during 1960-2001.

Table-19 : Agriculture: Work Force

Sl. No.	Year	% work force in agriculture and its allied activities
1.	1960	63.7
2.	1971	40.2
3.	1981	28.6
4.	1991	27.5
5.	2001	16.6

Source: Directorate of Agriculture.(2009-10)

From the trend it appears that the percentage of cultivators has declined from 63.7 in 1960 to 16.61 in 2001.

The contribution of this sector to the State Domestic Product has declined from 16.5 percentages in 1961 to below 8 percentages at present.

This declining trend towards agriculture could be attributed to small land holdings, high cost of wages and non availability of agricultural laborers to urban areas in search of especially white collar jobs. Goa today being scope for agriculture, high value added crops like horticulture, floriculture, medical plants, vegetables etc., with this scenario in view a number of measures have been imitated to revitalize the agriculture sector

The following table shows distribution of operational holding, according to the size class during 1990-91 and 2000-01.

Table-20 : Distribution of Operational Holding According to the Size Class during 1990-91 and 2000-01

Sl. No.	Size class	No. of Operational Holdings(1990-91)		No. of Operational Holdings(2000-01)	
		Total	Percentage	Total	Percentage
1.	Below 0.5	426999	61.00	37688	60.11
2.	0.5-1.0	142533	20.00	13266	21.16
3.	1.0-2.0	8340	12.00	6576	10.49
4.	2.0-3.0	2533	3.00	2330	3.72

Table Contd...

5. 3.0-4.0	1019	1.00	558	1.37
6. 4.0-5.0	532	1.00	558	0.89
7. 5.0-7.5	639	1.00	658	1.05
8. 7.5-10.00	357	-	242	0.39
9. 10.0-20.0	419	1.00	328	0.52
10. 20.0 & above	231	-	186	0.30
	71922	100	62694	100.00

Source: Economic Survey, 2008-09.

From the table, it reveals that there is a gradual decline in the number of operational holdings during the respective periods in different categories of holdings.

The following table shows the land use pattern of Goa 2007-08 in which different proportion of land is needed for various purposes.

Table-21 : Land Use Pattern of Goa 2007-08

Sl. No.	Item	Area in hectares	Percentage to total
1.	Total reporting area	361113	100.00
2.	Area under forest	125473	34.75
3.	Land not available for cultivation current fallow	37137	10.28
4.	Current fallow	9581	2.65
	Permanent pastures and other grazing land	1305	0.36
	Land under miscellaneous tree crops and groves not included in net area sown	580	0.16
	Cultivable waste including fallow land	52829	14.63
5	Net area sown	134208	37.17
6	Area sown more than once	35069	9.71
7	Gross cropped area	172108	47.66

GOA – MINERAL RESOURCES

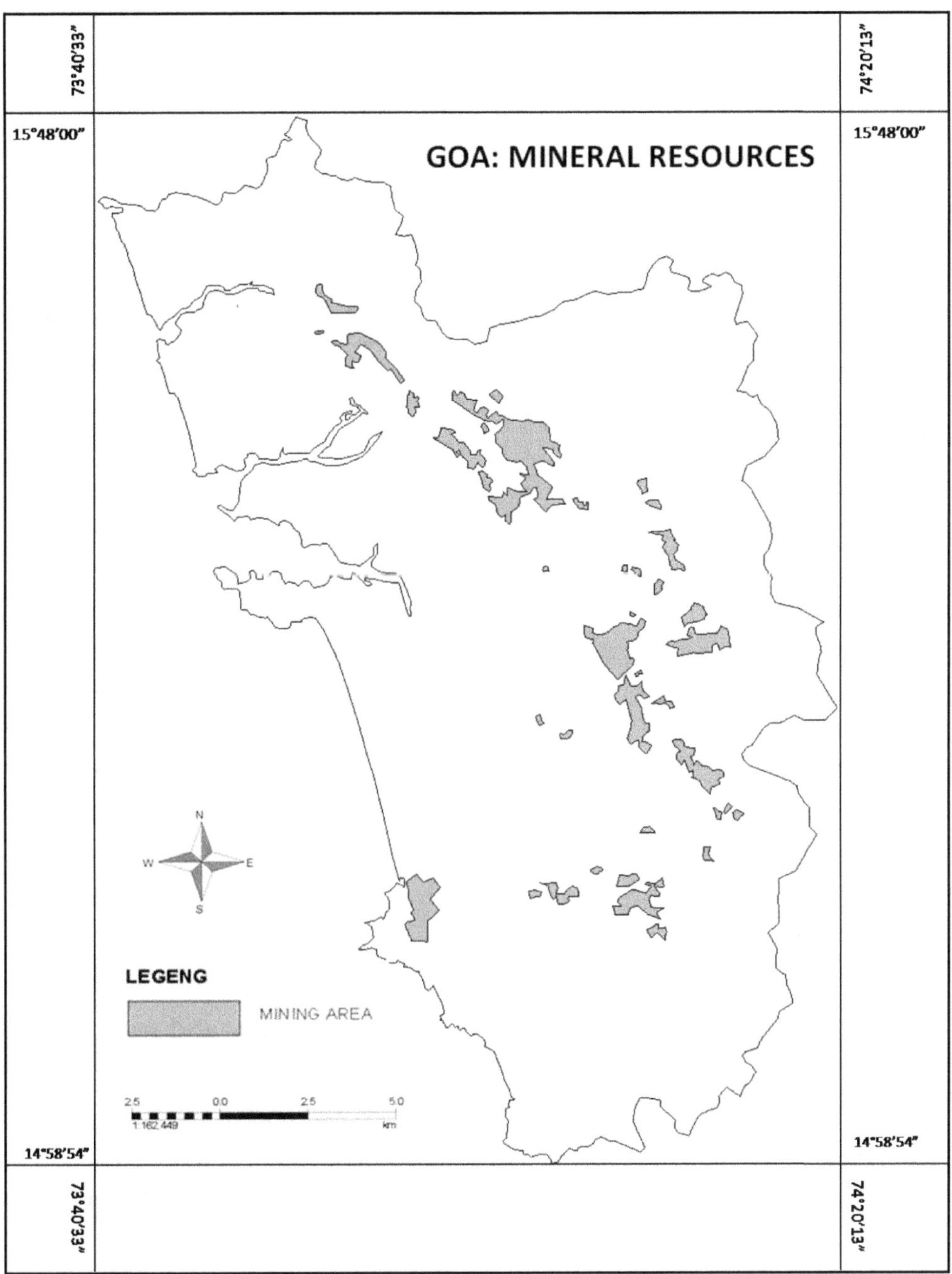

Map-8

As per the table total reported area is 361113 hectares, out of which 134208 Ha is net area sown, 35069 Ha is sown more than once and gross cropped has been reached to 1,72,108 hectares thus 47.66 of total land area is used for agricultural development.

The following table and graph depicts the production of agriculture and horticultural crops during the year 2003-04 to 2007-08.

Table-22 : Agricultural Production

Sl. No.	Crop	Units	Estimated production				
			2003-04	2004-05	2005-06	2006-07	2007-08
1	2	3	4	5	6	7	8
1.	Paddy	Ton	255974	217762	220853	193418	182519
2.	Ragi	"	394	337	463	254	170
3.	Maize	"	800	800	500	-	-
4.	Pulses	"	9147	9385	11570	16250	11261
5.	Groundnut	"	5673	5410	7942	4600	6998
6.	Sugarcane	"	57607	60583	55897	58279	56027
7.	Cashew nuts	"	23233	25556	27070	24380	21942
8.	Coconut	Nuts	122	123	125	127	128
9.	Areca nut	Ton	2650	2495	2578	2614	2666
10.	Mango	"	17800	18700	27075	19280	18894
11.	Banana	"	16900	19755	21026	23420	23450
12.	Pineapple	"	4500	4520	4515	5040	5544
13.	Vegetables	"	70467	74725	82580	84290	56027
14.	Other fruits	"	35100	38610	39304	39804	398

Source: Directorate of Agriculture, Government of Goa (2009).

There are various crops grown by the state. Paddy and horticulture, which are major crops best suited to favorable geographical conditions of are largely grown by the people of Goa.

The government is making sustained efforts to revitalize agriculture and allied activities in the state. The following are the important measures that have been initiated to promote agriculture and horticulture in Goa.

Table-23 : Production of Mineral Ores from 1998-99 to 2008-09. Metric tones in terms

Types of ore	1998 1999	1999 2000	2000 2001	2001 2002	2002 2003	2003 2004	2004 2005	2005 2006	2006 2007	2007 2008	2008 2009
Iron Ore	1552171	12172925	15862763	15737701	17371039	23727937	21705667	25440925	30738191	31327805	32720536
Manganese ore	17581	11706	9811	11946	8837	9433	7439	5773	3560	1454	1080
Bauxite	37170	32179	31950	58990	59694	81488	41934	28323	111,097	111259	463150

Source: Mineral Ore Exporters Association of Goa (2010)

MINING

Mining is the process of taking out minerals from the earth crust. Most Minerals extracted from the earth crust and a few from water also. Mineral is an inorganic substance, which is freely occurring in the nature. The significance of mining industry is rapidly increasing over the years in this modern age of machinery to meet growing needs of population of the world.

Prior to liberation of Goa from Portuguese rule, only a few small industrial units existed in the state (territory). The major economic activity was confined to the mining of Iron ore and manganese ore. Mining has been the important element in the economy history of modern Goa. It still plays an important role in employment generation and earning foreign exchange to the state. The state of Goa is richly endowed with Ore and Bauxite are the most explored minerals.

The mining belt of Goa covers approximately 700 sq.km almost 19% of total geographical area of the state mainly concentrated in Bicholin of North Goa district and Salcete, Sanrgue in and Quepem of South Goa district. Mining and associated activities have greatly affected the natural landscape in and around these areas which is characterized by the presence of pits and waste rejects. The mining belt of the Goa is divided into one namely, Northern central and

southern zone. Usages River is the dividing line between Northern and central zone, and Sanguem River between the central and southern zone. The maximum area under mining is in Sanguem taluka followed by Bicholim, Sattari and Pernem.

Iron ore deposite in Goa are mainly hematite and occupy really 1/5th of state major iron and belts are divided into four separate areas by major faults. Bicholim Sanvordem area, Pale area, Shirgao Kale area and Sanguem and Quepem area. Nearly 800 million tons of Iron deposits have been estimated in Goa. As far as bauxite resource is concerned, it has 7.17 million tons. Geological survey of India located few deposits occur 130 kilometers along the west coast i.e. Polem, Galgibag, Taligao, Bambolim, Pernem, Morjim and Coelim. The Best manganese ore deposits are also found in the central and Northern parts of Goa.

The mining industry in Goa started during the Portuguese rule period. However the process underwent modernization only after its independence. Most of the mines are in private hands. They were leased out by the Portuguese Authority in pre-liberation days.

Mining activity in 1950, Goa has grown in leaps and bounds, providing employment to many and contributing significantly to the nation in the form of foreign exchange. The mining provides an employment to the tune of 10,000 persons directly and 11,000 indirectly.

The following table shows the production of Iron, Manganese are and Bauxite during the period 1998-99-2008-09.

Table-24 : Export of Iron Ores from Goa (2004-05- 2008-09)

Year	Iron ores in million tonnes	Value in Rs (crores)
2004-05	24.46	2792.58
2005-06	27.03	3422.67
2006-07	33.49	4591.81
2007-08	32.23	6700.00
2008-09	36.00	7316.00

Source: Mineral Ore Exporters Association of Goa (2010).

The Iron ore export has increased from Rs. 18.31 crores in the year 1961 to Rs 7316.00 crores in the year 2008-09. More than 60% of the Iron ore exported from the country is from Goa alone. The value of ore exported from Goa is now more than the tax and non-tax revenue of the state, which indicates the importance of this industry to the economy importance of the state.

There is good demand for Goan mineral ore from European countries and Asian countries. Goa has been able to export sufficient quantities of Iron ore to the needy countries of the world like Japan, Netherlands, China, Belgium.

Role of Mormugao Port Trust (MPT) in Development of Mining in Goa

Mormugao port, a proto type Natural harbour is situated on the west coast of India in the state of Goa at the mouth of the river Zuari, MPT which handles roughly of the country is considered to be Iron ore loading port in India. The total iron ore export from Mormugao which stood at more lanel of 52.000 tons in the year 1946-47 rose to the peak lanel of 15.754 million tones in the year 1993-94. However this highest figure of the total Iron Ore exports from Mormugao surpassed 22.095 million tons in the year 2003-04. It further went up touching the highest export figure of 36.00 million tons in the year 2008-09 valued at Rs.7316/- crores earned by the state. Near before Goa had exported this amount of Iron one outside the country and earned good amount of foreign income.

Leading Mine Operators in Goa

1.	Sesa Goa	-	4,24,000 sq meters
2.	Chowgale Mines	-	2,20,000 sq meters
3.	Formento Mines Cr	-	1,20,000 sq meters
4.	Resource International	-	60,000 sq meters
5.	D.B.Bandodkar Mines	-	50,000 sq meters
6.	V.S.Dempo	-	30,000 sq meters
7.	Timblo Mines	-	20,000 sq meters
8.	Salagaonkar Mining Indutry	-	17,000 sq meters
9.	V.G.Quenim.	-	5,500 sq meters

INDUSTRY

Manufacturing industries are those industries which are concerned with the processing and altering of raw materials and semi finished products. The importance of manufacturing has become so great in the modern age that the economic strength of a country is measured in terms of its manufactures. Today every country or state is trying to develop its manufacturing industries more and more and this desire for the development of industries has led each country to search for suitable location within its borders.

Goa is a small state, which enjoys suitable climate, location and basic infrastructure facilities from the point of industrial development. Prior to the liberation of Goa, industrial development was not very good, but after the liberation there was gradual increase had growth of small scale industries and large medium scale industries took place in course of time.

The state of Goa envisages catalysing economic growth through accelerated industrial development. The mission is to create sustainable employment opportunities mainly to the people of Goa. It also includes environment friendly industrial development ensuring balanced growth of regions, a facilitative regime that explores and releases the energies of the private sector to create an environment in which industry both existing and new can prosper.

The state of Goa has a robust industrial base, contributing significantly to Goa's economy. As on 31-7-2009, the state has 8670 small scale units, of which 7093 were registered till September 2006 and the remaining 1407 units registered under EM part-I and 170 units under EM part-II upto April-June 2009. It shows commendable growth in the field of industry of Goa, employing more than 65089 persons. Investment in small scale industries has been growing rapidly over the years from Rs. 352.26 lakhs in the year 1970-71 to 61203.53 lakh in 2008-09, which in fact, clearly presents tremendous rise in the number of SSI units. Investment and employment for the people of Goa as well as outsiders thereby improving their living conditions and benefiting immensely Goa's economy.

The following table depicts the growth of small scale industries in Goa over a period of five decades from 1961-70 to April-June 2009, which is also depicted in the form of a graph.

The following table shows the growth of small scale units in the past seven years from April 2003 to March 2009.

The table reveals the fact that for the last seven years 524 units have been registered from 2003 to 2007 has gone up from 6714 to 7238 by 2009. There is considerable increase of SSI units due to the favourable condition and available sources in the study region.

GOA – INDUSTRIAL ESTATES

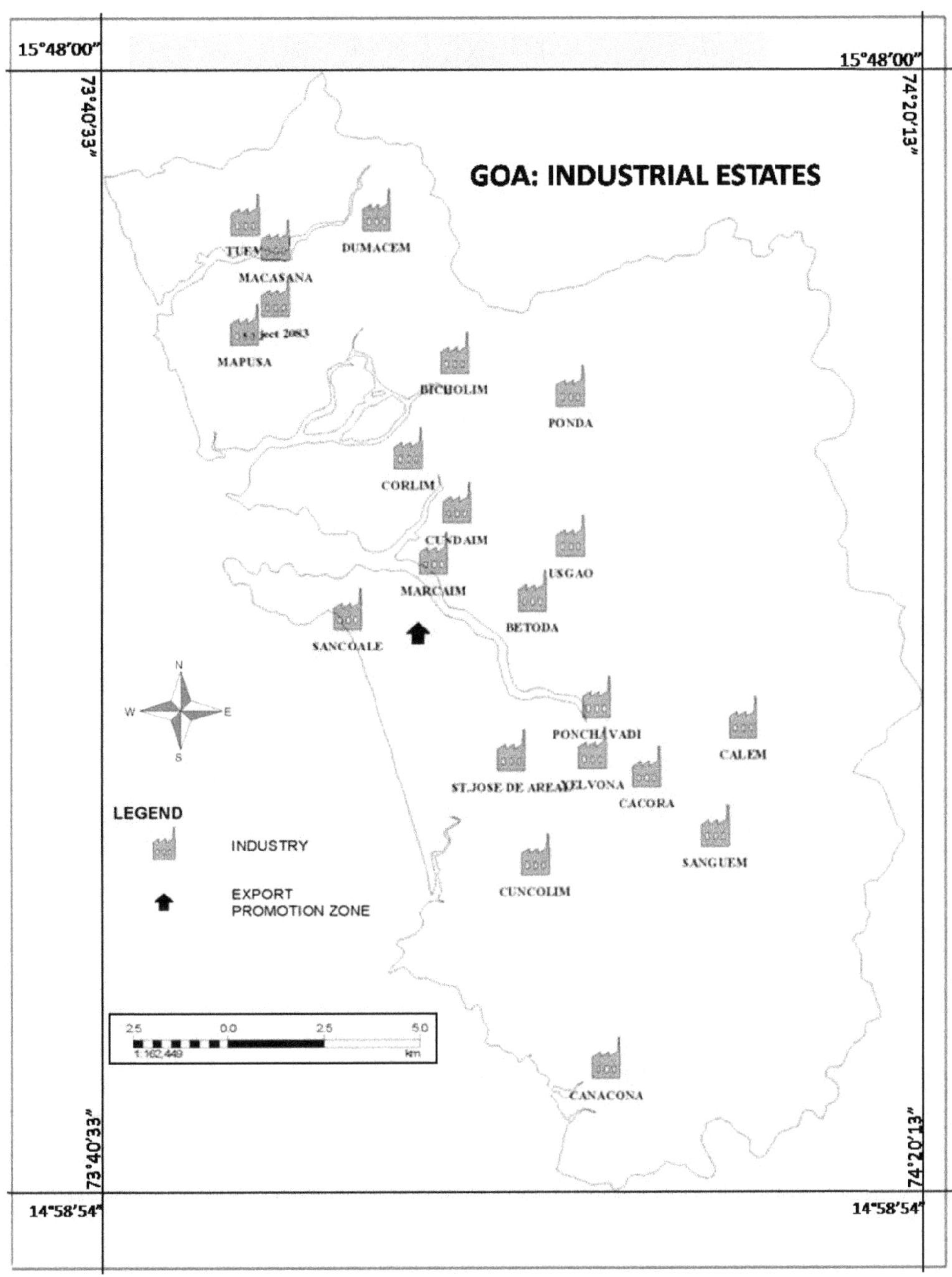

Map-9

Table-25 : Trend of SSI Units

Sl.No.	Year	No. SSI Units
1.	2003	6714
2.	2004	6852
3.	2005	6954
4.	2006	7060
5.	2007	7105
6.	2008	7162
7.	2009	7238

Source: Department of Industries and Mines, Government of Goa (2010).

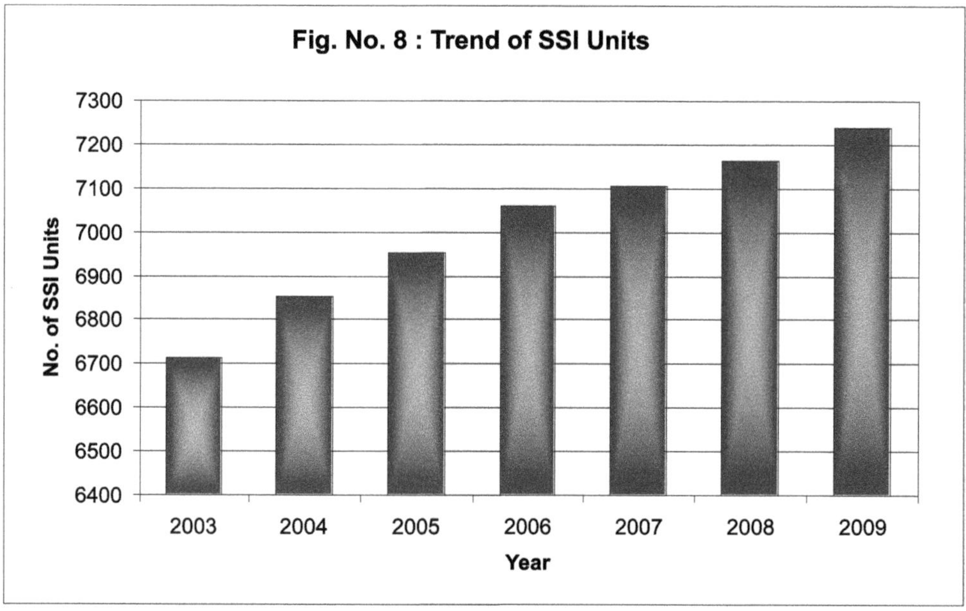

There have been continuous efforts to accelerate the pace of industrial development in the state. The new industrial policy was unveiled in the year 2003. The policy envisaged various like capital contribution scheme, share capital to local entrepreneurs and self-employed, preferential purchase incentives for SSI interest subsidy scheme, etc. This has lead to greater industrial development in Goa.

INDUSTRIAL ESTATES

The Goa Industrial Development Corporation has established 20 industrial estates in the state covering all talukas. The corporation has allotted

2050 plots to these estates, of which 150 are vacant. So far, 716 sheds have been allotted wherein 1128 units are functioning. Infrastructure facilities such as water supply system, asphalted roads, street lights, drainage, canteen, post office, banking facilities, warehousing facilities and other need based requirements, including police outpost have been created.

Table-26 : Goa – Distribution of Industrial Estates

Sl.No.	Estates	Plots Allotted	Units Functioning
1.	CORLIM	79	58
2.	MARGAO	37	80
3.	SANCOALE	160	81
4.	MAPUSA	17	37
5.	TIVIM	64	95
6.	BICHOLIM	82	70
7.	HONDA	26	24
8.	BETHORA	86	80
9.	KUNDIAM	339	172
10.	CANACONA	68	24
11.	TUEM	62	29
12.	KAKODA	65	39
13.	VERNA	433	184
14.	CUNCOLIM	218	57
15.	PILERNE	130	46
16.	MARCAIM	116	43
17.	SHIRODA	-	-
18.	COLVALE	24	4
19.	PISSURLEM	43	5
20.	SANGUEM	1	-

Source: Eco. Survey, Directorate of Planning, Statistics and Evaluation, 2004-05.

Goa has taken the initiative to implement infrastructure development projects. Power, water supply and transport sectors are the key areas identified for infrastructure development. The Government has floated the Goa State Infrastructure Development Corporation for supply for speedy implementation of infrastructure project.

LARGE AND MEDIUM SCALE INDUSTRIES OF GOA

Manufacturing industries are those involved in alteration of raw materials into variety of semi-finished goods into finished goods through various manufacturing processes. The more you alter or processes the raw material, the more valuable it becomes. Thus manufacturing processes is essential in order to get wide range of finished goods for our direct consumption or various purposes.

Apart from small scale industrial development in Goa, the attempt has been made to understand the scenario of large and medium scale industries in the study area. There are various factors influencing the location and development of large and medium scale industries all over Goa, *i.e.*, availability of land, continuous favourable climate, pawn supply, labour, finance, efficient industrial policy (GIDC) EDC, etc. Transport, road, rail, water and availability of port facilities, i.e., MPT, etc.

The following table shows the trend of large and medium scale industries development in Goa during 1960-2009.

Table-27 : Large and Medium Scale Industrial Development

Year	No. of Large Scale Units	Employment	Investment in Crores (Rs.)
1960	1	168	0.76
1970	9	3849	19.64
1974	13	1899	137.16
1983	26	8103	325.30
1991	44	9245	576.30
1995	71	12596	732.93
1999	140	18896	1068.73
2003	150	22789	5159.63
2009	172	26632	6300.55

Source: Department of Industries and Mines, Government of Goa, 2009

From the table 2.26 given, it has been observed that the study area has great potentiality to promote and develop large and medium scale industries. It was as slow as to begin with single large scale unit at Zuari Nagar in Marmugoa taluk, *i.e.*, Zuari Agro Chemicals Limited established in the year 1960, with the employment capacity of 168 persons and investment of Rs. 0.76 crores. It was just the beginning of large scale industrial development in the study area. The first unit has produced a variety of agricultural products, i.e., fertilizers, urea, phosphate, sulphate, etc. Further, the number of units increased to 9 in 1970, followed by 26 in 1983, 44 in 1991 and 71 in 1995 respectively. Wide range of manufacturing industries have flourished in Goa. The prominent ores i.e., D-Link, MRF tyres, Ciba Giegy Ltd., Nestle, Aventis, Cipla, Glen Mark, Zandu pharmaceuticals, Lupin, Binany Glass and Fibers Ltd. TCS, Phil Corporation, Goa Telecommunications, G.K. Industries, Sanjivani Sugar Factory, Goa Shipyard Ltd., etc. During 1995-2003, rapid growth of large scale industries took place. The number of units increased from 71 in 1995 to 150 in the year 2003. Within a span of eight years, additional 79 units have come up, which indicates more than 100 percent growth has taken place. Therefore, this period could be termed 'Industrialization" in true sense. The whole credit goes to GIDC, EDC, Government of Goa, for their continuous support and encouragement towards promotion of industrial development in the study area through various schemes and plans. On this account, the state of Goa can be proud of itself to become one of important industrial states in the country catering to the needs of local and international demand.

POWER SECTOR OF GOA

The power sector is the heart of the state's economy whether in mining, tourism, industry or agriculture or any other priority sector. Power is the primary catalyst for growth. Neglect power, and the whole economy will suffer. Goa has one 48 MW naphtha based power generation plant in the private sector. The state is mostly dependent on the power allocated from the neighbouring state grids.

Table-28 : Peek Demand for the Year 2008-09

Recorded demand from Western Region	234MW
Recorded demand from southern region	69MW
Total recorded demand from central grid	312MW
Recorded demand from REL to GOG grid	16MW
Recorded demand from REL to direct consumers	26MW
Total demand	354MW

Source: Goa Infrastructure Report-GCCI (2009).

The following are the important measures taken by the government to ensure better power supply and other related services.

- The government has decided to award the execution of some of the major EHV works on deposit to POWERGRID. Accordingly, the complete renovation and modernization of the important 220/110/33 KV sub-station at Ponda (Rs. 42.00 crore), the work of conversion from overhead to underground network in the balanced areas of Margao (Rs. 81.00 crore) is being awarded to POWERGRID.
- The government took a decision to levy electricity duty and set aside the amount so collected in 'ELECTRICITY DUTY FUND' to be exclusively utilized for development of the transmission and sub-transmission network in the state. During the year 2008-09, an amount of around Rs. 50.00 crore is expected to be set aside in this fund.
- One time settlement scheme for defaulting consumers of electricity having accumulated arrears upto Rs. 10,000/-, had been announced in the budget 2008-09. The scheme has been modified to the extent of accumulated arrears of Rs. 20, 000/- and has been recently notified.
- Notification in respect of charging electricity duty of 18 paisa to domestic and agricultural consumers and 58 paisa to industrial and commercial consumers to accelerate infrastructure development in the power sector has been notified in May 2008.
- The state electrical inspectorate has become functional during the year 2008-09. Goa has also joined the joint Electricity Regulatory Commission, which has now become functional. The future tariff revision will be subject to approval of the JERC, for which the Power Department will have to file tariff petitions.
- By strict implementation of the financial reduction plan through the third supplementary agreement executed with RSPCL, the Government has considerably reduced the financial burden on drawn of power from RSPCL. A further plan to reduce the financial burden, to some extent, has been worked out.
- A Billing Dispute Redressal Committee has been constituted for settling billing disputes, and has become operational, which enables a consumer to settle the disputed bill and the Department, in turn gets the revenue.

- State of the art electronic meters are being provided to new consumers. Besides, a drive has been undertaken for replacement of non working meters by state of the art meters.
- The Department has also provided operation and maintenance division with trolley mounted mobile transformers to enable the power supply to be restored within a couple of hours to the locality in case of breakdown of the distribution transformer centres even at night. This greatly improves reliability of power supply.
- Implementation of the accelerated power development and reforms programme (APDRP) of the ministry of power:
    - The sub-transmission and distribution improvement works taken up in the state in the first phase are on the verge of completion.
    - The underground conversion works in Margao and Panaji have been completed.

BANKING

The role of the banks in economic development is to remove the deficiency of capital by stimulating savings and investment. The bank encourages the habit of saving among the people and enables small savings to be accumulated into large funds, thus available for investment.

Goa has achieved a very high level of socio-economic development in terms of low poverty, good standard of living, pollution control, low birth rate, etc. Reflection of these achievements is visible in the banking sector in terms of growth of banking offices, number of banks accounts and high rare of saving, etc. As on date, there are 510 banking offices in Goa, out of which 303 are in north District and 207 are in South District.

The number of banking offices of different banks is given in the following table.

Table-29 : District wise Distribution of Banks in Goa

	North Goa	South Goa	Total
Nationalized Banks	187	133	320
Co-Operative Banks	77	48	125
Private Sector	34	26	60
Financial Bank	05	-	05
Total	303	207	510

Source: Economic Survey (2008-09).

Goa has a very high rate of savings. The deposits in banks have been increased to Rs. 25596.25 crore till March 31st 2009. An important characteristic of deposits in Goa is the significant contribution from Non Resident Emigrants. NRI deposits have reached Rs. 4519 crore for the same year.

TOURISM IN GOA

Tourism plays an important role in the economy of Goa, for Goa, tourism generates 13.7% of the state's Net Domestic Product, 7% of tax revenue. Goa is capitalized on its unique historical and cultural heritage. Tourism in Goa has assumed the role of major economic activity, having direct and indirect correlation with all other economic sectors. Goa is endowed with rich natural environment, historical and cultural heritage. Tourism has been major factor for promoting development in Goa.

Goa is a tiny emerald land, situated well on the Western coast of India. Goa is a state with beautiful coastline of 104 kms long, surrounded by mountains, rivers, lakes, back waters and other tourist destinations. Besides a harmonious blend of Eastern and Western cultures, very friendly and hospitable people make, Goa a veritable tourist paradise. Goa is indeed, a place of tranquility and most sought after state tourist destination for the visit of domestic as well as tourists from all over the world.

Goa is a treasure of heritage, centuries old temples, churches forts monuments, heritage houses, etc. Tourism has contributed substantially to the economic development of the state by way of foreign exchange earnings. Employment generation, intra sectoral competition, in turn, improved the living standard of the people. Beautification of important tourist destinations, improvements of roads in tourist areas and improvement of life safety measures, regular sight seeing tours, cruises, tour packages, easy access to accommodation, etc., have caused significant tourist flow in Goa. Tourism is being diversified to hinterland by promoting heritage tourism, village tourism, eco tourism, health tourism, adventure tourism, etc.

Tourism took birth in Goa in the 60's, but its developments began only in the 80's, and have reached today the position of competing with other tourist destinations in the world. Although, Goa joined mainstream only after 14 years of the country's independence, tourist traffic to Goa registered a phenomenal growth from 200 lakh people in 1975 to 24,48,959 in 2004. It further went up to 26 lakhs in the year 2007. Goa's income from tourism in terms of foreign exchange earned has gone up from Rs. 32.64 crores in

1986-87, to approximately Rs. 600 crores in 2001 which represents a steady rise of about 25% every year.

The number of tourists visiting Goa is growing at rapid rate. In fact, the state has already surpassed the tourist arrival projection made for the year 2016. From 12,68,513 in the year 2000, the number of tourist arriving in Goa increased to 23,71,539 in 2008 registering on compound annual growth rate of 17% as against the assumption of 3.67 made by the tourism master plan 2011. From 2003-2008, domestic tourist arrivals have witnessed an increase of 14.6% foreign tourist.

The following table shows the tourist arrivals in the state during the period 1985-2008.

Table-30 : Tourist Arrivals in the State

Year	Domestic	Foreign	Total	Percentage
1985	682545	92667	775212	-
1986	736548	97533	834081	7.6
1987	766846	94602	861448	3.3
1988	761859	93076	854935	-0.7
1989	771013	91430	862443	0.9
1990	776993	104330	881323	2.2
1991	756786	78281	835067	-5.6
1992	774568	121442	896010	7.3
1993	798576	170658	969234	8.2
1994	849404	210191	1059595	9.3
1995	878487	229218	1107705	4.5
1996	888914	237216	1126130	1.7
1997	928925	261673	1190598	5.7
1998	953212	275047	1228259	3.2
1999	960114	284298	124412	1.3
2000	976804	291709	1268513	1.9
2001	1120242	260071	1380313	8.8
2002	1325296	271645	1596941	15.7
2003	1725140	314357	2039497	27.7
2004	2085729	363230	2448959	20.1
2005	1965343	336803	2302146	-6.0
2006	2098654	380414	2479068	7.7
2007	2208986	388457	2597443	4.8
2008	2020416	351123	2371539	-8.7

The following table presents nationality wise break up of foreign tourist from 2000-07.

Table-31 : Nationality wise Break-up of Foreign Tourist

Sl.No.	Countries	No. of Tourist Arrived	Percentage
1.	U.K	151123	38.90
2.	RUSSIA	36927	9.51
3.	GERMANY	30768	7.29
4.	FINLAND	23583	6.07
5.	FRANCE	16659	4.29
6.	SWTZERLAND	11840	3.05
7.	SWDEN	14543	3.74
8.	U.S.A	8827	2.27
9.	AUSTRALIA	7621	1.96
10.	SOUTH AFRICA	3383	0.87
11.	BRAZIL	2387	0.61
12.	ITALY	2429	0.88
13.	CANADA	3364	0.87

14.	JAPAN	2258	0.58
15.	DENMARK	2321	0.60
16	AUSTRIA	1648	0.42
17.	HOLAND	15930	0.41
18.	PORTUGAL	1234	0.32
19.	IRELAND	1398	0.36
20.	BELGIUM	306	0.08
21.	NORWAY	319	0.08
22.	IRAN	926	0.24
23.	U.A.E	721	0.19
24.	NEW ZELAND	216	0.06
25.	GREEK	64	0.02
26.	OTHERS	60999	15.07
	TOTAL	388457	100

Source: Department of Tourism, Government of Goa (2008).

Form the above table it appears that foreign tourists are coming from all over the world, i.e., cold climate European, North America, Latin America, Africa and Asian countries.

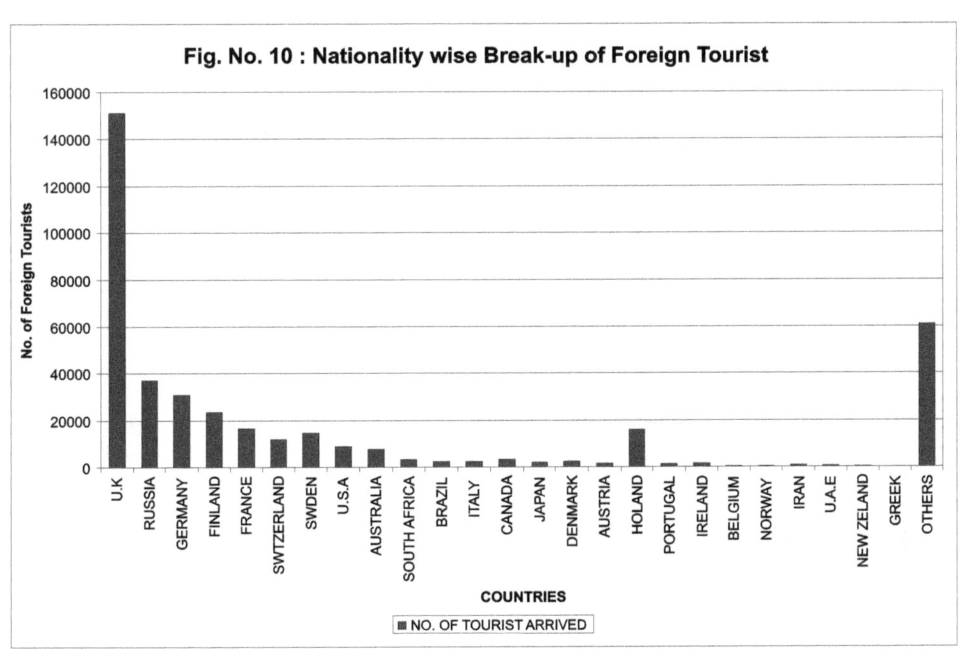

Table-32 : Nationality wise Break-up of Foreign Tourists from 2000-07

Sl.No.	Countires	2000	2001	2002	2003	2004	2005	2006	2007
1	U.K	94638	90756	96242	125623	149998	144672	158447	151123
2.	RUSSIA	2363	1892	6602	12975	25613	29473	32293	36927
3.	GERMANY	28924	17823	16798	21715	26510	19743	23654	30768
4.	FINLAND	11492	7898	10832	36897	35123	18811	25724	23583
5.	FRANCE	11271	9095	9364	6489	8069	12574	15336	16659
6.	SWTZERLAND	10870	8708	8949	6453	9448	10763	13421	11840
7.	SWDEN	6532	8341	8482	8967	10826	7626	9954	14543
8.	U.S.A	10326	5978	5787	4158	6110	6175	8303	8824
9.	AUSTRALIA	4697	3696	3493	2611	3022	5073	7482	7621
10.	SOUTH AFRICA	3812	2467	1769	1515	0	4376	5238	3383
11.	BRAZIL	0	0	0	0	0	3237	3435	2387
12.	ITALY	10136	8567	7236	3364	3832	3216	3397	3429
13.	CANADA	4518	4620	4788	3246	3040	2978	3124	3364
14.	JAPAN	3645	2129	2294	0	0	1934	2163	2258
15.	DENMARK	3494	3897	4091	2361	2550	1852	2012	2321
16	AUSTRIA	4103	3972	3637	2451	2504	1698	1827	1648
17.	HOLAND	3588	3938	4121	2863	4182	1361	1447	1593
18.	PORTUGAL	5216	5461	5153	5296	3839	1232	1487	1234
19.	IRELAND	0	2873	2163	1905	1668	1135	1243	1398
20.	BELGIUM	0	0	0	0	2226	0	86	306

Contd...

Table-32 Contd...

21.	NORWAY	0	0	0	1061	1110	0	93	319
22.	IRAN	0	0	0	0	2347	0	68	926
23.	U.A.E	0	0	0	0	0	0	0	721
24.	NEW ZELAND	4056	3151	2684	1172	0	0	177	216
25.	GREEK	3198	0	0	0	0	0	128	64
26.	OTHERS	64830	64809	67160	63235	61213	58814	56869	60999
	TOTAL	291709	260071	271645	314357	363230	336803	380414	388457

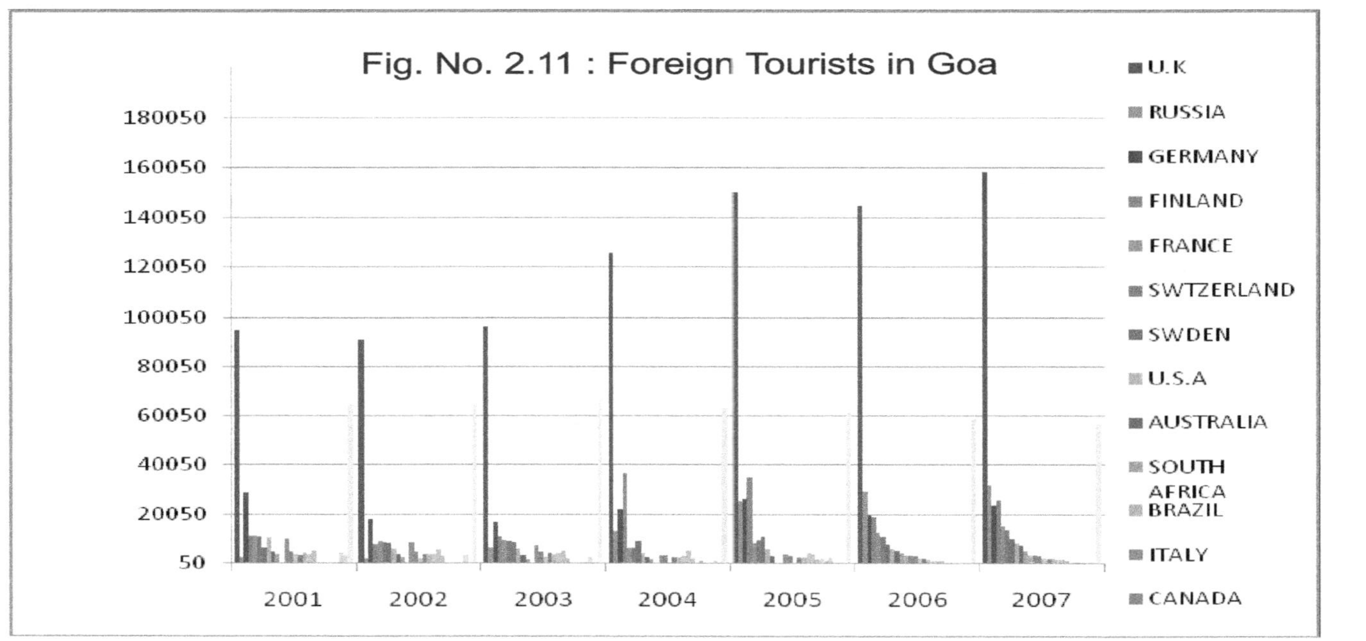

Fig. No. 2.11 : Foreign Tourists in Goa

The number of charter flights arriving in Goa has also increased. Nearly 50% of the foreign tourist arrivals are through charter flights. The following table and graph represents tourist arrival by foreign charter flights during 1985-86 to 2007-08.

Table-33 : Tourist Arrival by Foreign Charter Flights

Year	No. of Flights	Passengers
1985-86	24	3568
1986-87	26	4401
1987-88	25	5419
1988-89	83	9705
1989-90	107	9266
1990-91	41	5815
1991-92	121	17102
1992-93	259	39871
1993-94	299	58369
1994-95	313	59881
1995-96	337	75694
1996-97	282	73172
1997-98	350	88817
1998-99	356	90635
1999-00	405	94289
2000-01	419	116992
2001-02	279	76410
2002-03	384	94350
2003-04	532	126255
2004-05	690	158993
2005-06	719	180310
2006-07	720	169836
2007-08	710	175951

Source: Department of Tourism, Government of Goa (2008).

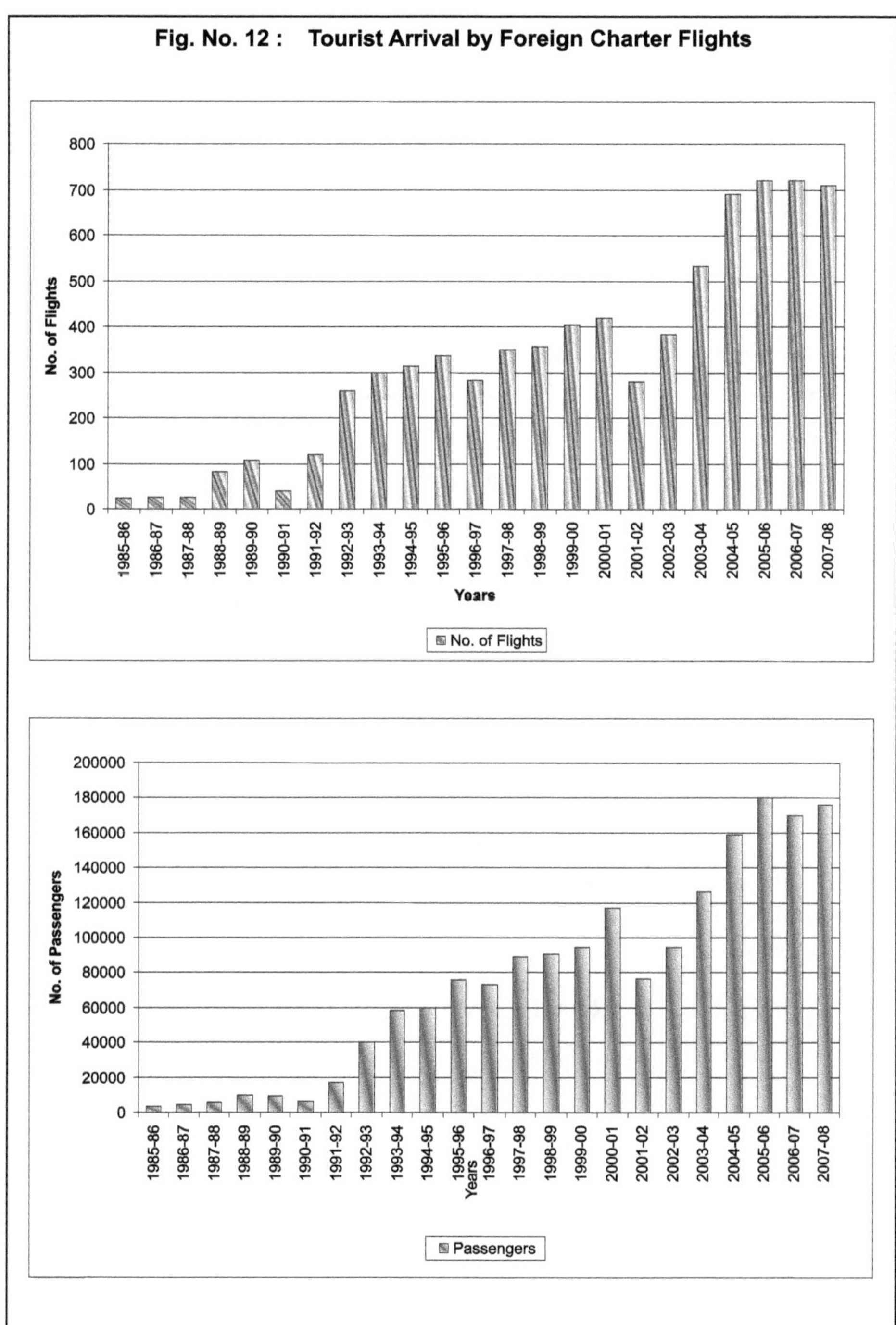
Fig. No. 12 : Tourist Arrival by Foreign Charter Flights
800
700
600
500
400
300
200
100
0
No. of Flights
1985-86
1986-87
1987-88
1988-89
1989-90
1990-91
1991-92
1992-93
1993-94
1994-95
1995-96
1996-97
1997-98
1998-99
1999-00
2000-01
2001-02
2002-03
2003-04
2004-05
2005-06
2006-07
2007-08
Years
No. of Flights
200000
180000
160000
140000
120000
100000
80000
60000
40000
20000
0
No. of Passengers
1985-86
1986-87
1987-88
1988-89
1989-90
1990-91
1991-92
1992-93
1993-94
1994-95
1995-96
1996-97
1997-98
1998-99
1999-00
2000-01
2001-02
2002-03
2003-04
2004-05
2005-06
2006-07
2007-08
Years
Passengers

From the above table and graphical representation, it appears that charter flights has been increased from 24 in 1985 to 710 in 2007-08. Similarly, the number of passengers has gone up from 3568 in 1985-86 to 175951 during 2007-08, which shows a rapid growth in the number of passengers coming from all over the world by charter flights over the years due to attractive tourist proposals in Goa. In addition to foreign tourists, more domestic tourists are also choosing to fly to Goa. During peak months (October to February) 5 to 6 charter arrives at Goa daily.

Transport Sector in Goa

The state has excellent road network. Road connectivity within the state amounts to more than 2000 km of road per 1000 sq.km of area as against the National average of 650.

In Goa there are three main national highways N.H.4A, N.H.17, N.H.17A. Panaji, the capital city of Goa is connected by N.H.4A from Belgaum in Karnataka. The N.H.17 starts in Mahad in Maharashtra state and enters in Goa through Patradevi and passes through talukas Pernem, Bardez, Tiswadi, Salcete and Canacona talukas. The third highway, N.H.17A is from Cortalim to Mormugao harbour. Cities Panaji and Margao are well connected by road from Mumbai, Pune, Kolhapur, Miraj cities in Maharashtra and Bangalore, Belgaum, Hubli cities in Karnataka.

Besides the National Highway, a dense network of metallic roads also connects the state to other parts of the country. Goa is connected by road to all major towns in India. Services are available to Mumbai, Pune, Bangalore, Kolhapur and Hyderabad. A well spread internal network of roads connects places within the state.

Table-34 : Road Development in Goa State

Sl.No.	Type of Road	Length (kms)
1.	National Highways	264
2.	State Highways	226
3.	Major District Roads	797
4.	Village Roads	2969

Source: AEZ Report (2004-05)

Goa's state owned Kadamba Transport Corporation and Private sector transport services connect interior villages to urban centers.

Road Transport

Road transport plays a vital role in socio-economic development of Goa to a great extent. The number of motor vehicles registered in the state stands at 672664 in the year March 2009. On an average, over 50,000 vehicles are being registered every year. The number of vehicles registered during 2008-09 is 49435. The distribution of vehicles by type is given in annexure 30. It is evident that about 71% of vehicles are in the category of two wheelers, followed by cars and jeeps, including taxis (about 20%). The number of vehicles registered during the period 1999-00 to 2008-09 is given in table no 35 and 36 respectively.

Smart Card Driving License

The department provides easy and efficient services to public in registration of new vehicles through authorized dealers and issue of Smart Card driving license which is durable, elegant and easy to carry. These services have been commissioned at RTO Panaji, Margao, Mapusa, Vasco, Bicholim, Quepem and Ponda. Under this project, 202079 people have been issued the Smart Card driving licenses up to 31-3-09.

Subsidy for purchase of yellow/black motor cycles/auto rickshaws/taxi.

The government has continued the implementation of the scheme for grant of subsidy for replacement of buses and minibuses older than 15 years, in order to curtail vehicular pollution and to promote clean environment. Similarly, the department has launched a new subsidy scheme for replacement and procurement of new yellow/black motorcycles, yellow/ black auto rickshaws and yellow/ black taxis in order to support this small scale self employed vulnerable section of the society and to ensure that old, unsafe and pollution generating passenger transport vehicles are removed from the road to protect environment and to enhance road safety. Under the scheme, during the year 2008-09, Rs. 23.97 lakh has been disbursed.

The department has introduced upgraded schemes of Pollution under Control (PUC) Certified status. Under the scheme, all stations authorized to issue PUC certificates for vehicles have been upgraded with modern computerization equipment facilities to test the pollution level of a vehicle. So far, 28 PUC centres have been issued license under the scheme.

Table-35 : Distribution of vehicles by type for the year 2006-07, 2007-08 and 2008-09

Sl. No.	Type of vehicle	2006 2007	2007 2008	M.V. on register as on 31-3-08 (cumulative)	2008 2009	M.V. register as on 31-3-09 (cumulative)	Percentage to the total
1	2	3	4	5	6	7	8
1.	Motor cycles on hire	1543	969	9050	2154	11204	1.67
2.	Goods vehicles	3297	2542	39882	2843	42725	6.35
3.	Taxis	880	1644	11885	867	12752	1.90
4.	Buses, mini buses and KTC	687	533	6909	706	7615	1.13
5.	Auto rickshaws	89	21	3607	51	3658	0.54
	Transport total	6496	5709	71333	6621	77954	11.59
6.	Motor cycles and scooters	32715	27852	436147	31145	467292	69.47
7.	Private cares and jeeps	10001	10261	109809	11050	120859	17.97
8.	Tractors/others	396	518	1647	507	2154	0.32
9.	Government vehicles		141	112	4293	112	4405 0.65
	Non- transport total	43256	98749	551896	42814	594710	88.41
	Grand total	49752	44458	623229	49435	672664	100.00

Source: Directorate of Transport, Government of Goa (2010).

Table-36 : Number of new vehicles registered from 1999-2000 to 2008-09 (up to December 2008)

Sl.No.	Year	No. of Vehicles
1.	1999-00	29869
2.	2000-01	25779
3.	2001-02	27189
4.	2002-03	33106
5.	2003-04	40050
6.	2004-05	46183
7.	2005-06	46716
8.	2006-07	49752
9.	2007-08	44403
10.	2008-09	49435

Source: Directorate of Transport, Government of Goa (2010).

Rail links

Rail network helps the flow of traffic both passengers and goods from Goa to various parts of India. The state is connected by Konkan railway to

Mumbai, Mangalore and Kerala. The conversion of Vasco-Miraj into broad gauge has facilitated direct rail journey from Goa to New Delhi. The state is connected through south central railway to Bangalore, Chennai and Hyderabad and New Delhi. There has been increase in Rail frequency on the routes over the years.

Air transport

Goa has a well developed international airport with facilities from charter flights and cargo at Dabolim. Goa is connected by air line service at the airport Dabolim to the rest of the country as well as abroad. The airport is about 30kms from t he capital city of Panaji, on the coast near Vasco-D-Gama, and is owned by Indian Navy. Most domestic airlines operate from here, apart from the chartered private airlines operating from UK and Germany. Dabolim airport is not an international airport, and therefore, there are no direct international flights to Goa. However, most of the major cities of India are connected to Goa through air. The cities of Delhi, Mumbai, and Chennai are connected internationally. There are many flights, which arrive daily in Delhi, Mumbai and Chennai, via which one can take a flight to Goa.

There is no dearth of flights for reaching Goa. Goa cheap flights, Goa discount flights, Goa chartered flights and a number of other flights are available for which online booking can be done. Last minute flights are also available.

Major airline operator like Indian, Jet Airways, Air Sahara, Spice Jet and Air Deccan, etc., has flights operating in and out of Goa. There are also regular flights between Dabolim Airport and Ahmedabad, Hyderabad, Kolkata, Bangalore, Udaipur, Thiruvananthapuram and Kochi.

There are many international flights, which carry passengers to Goa via other Indian cities. One way to come to Goa is by taking a flight to Mumbai, which is the major commercial hub of India. Flights like British Airways and Air India operate flights via Mumbai. Besides, Emirates via Middle East as well as Gulf Air, Kuwait Airways and Royal Jordanian also have flights via Mumbai. The national airline, Air India also lands a few international flights, especially from the Gulf countries.

For tourist from foreign countries, who do not want to take routes via other places, there are direct holiday charter flights from the UK (Manchester and Gatwick), Scandinavia, Holland and Germany. From an international destination it is approximately 10 to 11 hours. Several charter companies fly in to Goa's Dabolim Airport, direct from UK and other European countries during winter.

Direct charter flights from Holland, Switzerland and Scandinavia also operate to Goa. Air France, Lufthansa and Swiss Air also carry passengers to Goa via Mumbai. Besides, Condor is supposed to start its maiden weekly flights from Germany in October-November this year. Delhi and Mumbai have direct flights to Goa daily. There are two flights per week via Delhi and Mumbai. Almost all the international flights arrive in Mumbai in the late night. From Mumbai it is possible to take a domestic flight to Goa, in the early morning. It takes around 1 hour to reach Goa from Mumbai. Besides, flights like Indian and Air Sahara, connects Goa with Delhi. It takes approximately two and an half hour from Delhi to Goa. From Chennai it takes around 3 hours and from Bangalore it takes approximately one and an half hour to reach Goa.

Most of the major international airlines usually have agreements with one or more of their domestic counterparts in India, so in case of somebody arriving from abroad, it is possible to book the onward domestic flight at the same time as the payment is made for the international ticket. Most of the local airlines have a contact phone number in the city as well as at the airport where the latest information about flights status is always available. In the year 2006, Goa is expected to attract an increasing number of foreign tourists. It anyway is the most popular tourist destination in the Asia- Pacific region. This tourist paradise is on holiday twelve months a year.

Goa is a land not only of beauty and fun; it is also equipped to handle emergencies. Given below is a compiled list of the important emergency numbers for air transport in Goa that you might need:

Air port Authorities of India Dabolim: 512788

Air India: 513445

Indian Airlines: 513863

Jet Air ways: 510354

International flights to Gulf countries have been introduced. International charter flights land in Goa from different countries like UK, Germany, Russia, Finland, Sweden and Norway.

GOA – LOCATION OF PORTS

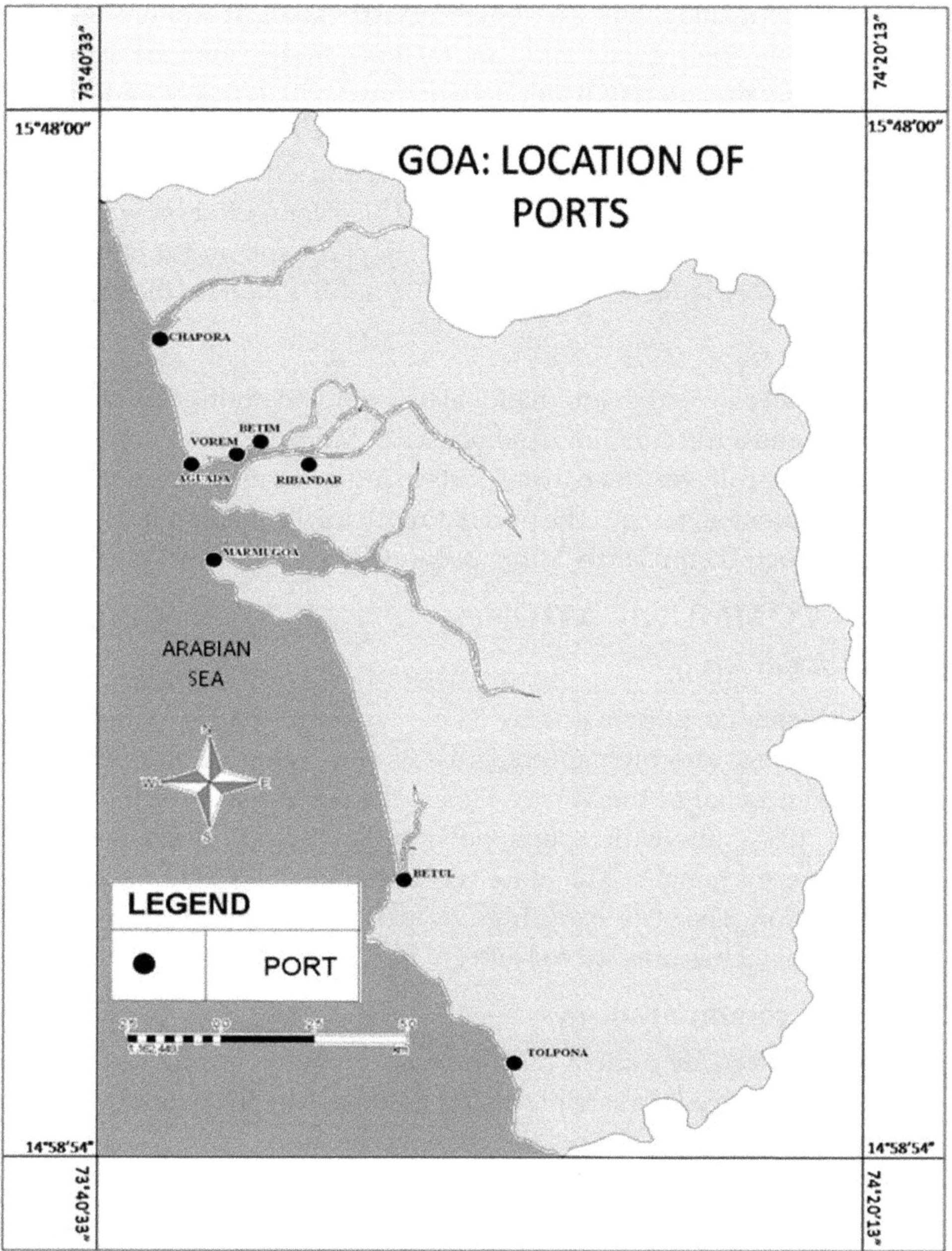

Map-11

Water transport

Goa enjoys a well developed internal water transport network formed by a grid of navigable rivers. This offers industries a most economical mode of transport for their goods and raw material throughout the state. The barges that ply the internal water network are swift, efficient and reliable. Most of the rivers in Goa are used for waterways. Ferry boats were means of crossing rivers in the state. Rivers Mandovi and Zuari are being used for carrying ore to Mormugao harbour. Goa is also connected by water ways from Bombay. Goa has a river navigation department, and it has 48 ferry boats and 2 motor launches to provide ferry service all over the state.

Port

Goa has an all weather international harbour and multipurpose general cargo berth and a full containerized service at Mormugao. Mormugao port is one of the best all weather natural ports in India. It can accommodate over 50 ships in outer anchorage. The port has mechanized loading facility, an oil berth and general cargo berth. Minor ports are also available along the river.

MEDIA COMMUNICATION

Telecommunication

The telecom facilities in goa are on par with metro cities in the country. Optical fibre provides high speed access to wide range of internet related services from Email to the WWW. Goa is the second state of the country to achieve 100% automatic telephone system, with a very good network of telephone exchanges. The state has excellent ISD/STD network for communication. The state has a high teledensity of 12 telephones per 100 inhabitants, which is only 2nd to Delhi.

Media and communication

Goa is served by almost all television channels available in India. Channels are received via satellite dishes. Doordashan, the nationl television broadcaster, has two free terrestrial channels on air.

DTH (direct to home) TV services are available from Dish TV, Tata sky and DD Direct plus. The All India Radio is the only radio channel in the state, broadcasting in both FM and AM bands. Two AM channels are broadcast, the primary channel at 1287 kHz and the VIvidh Bharati channel at 1539 kHz. AIR's FM channel is called FM Rainbow and is broadcast at 105.4 MHz.

private FM radio channels available are Big FM at 92.7 MHz, Radio Mirchi at 98.3 MHz, and Radio Indigo at 91.9 MHz. there is also an educational radio channel, Gyan Vani, run by IGNOYU broadcast from Panji at 107.8 MHz. in 2006, St Xavier's College, Mapusa, became the first college in the state to launch a campus community radio station 'Voice of Xavier's'.

Major cellular services operate include Reliance Infocomm, Tata Indicom, Vodafone (previously Hutch), Bharti Airtel, BSNL and Idea cellular.

Local newspaper publications include the English language the Herald (goa's oldest, once a Portuguse language paper known as Oheraldo), the Gomantak Times and the Navhind Times. In addition to these, The Times of India and the Indian Express are also received from Bombay and Bangalore in the urban areas. The Times of India has recently started publication from Goa itself serving the local population news directly from the state capital. Among the list of officially accredited newspapers are The Navhind Times, The Herald Times and The Gomantak Times (all in the English language) and Gomantak, Tarun Bharat, Navprabha, Pudhari, Goa Times, Sanatan Prabhat, Govandoot (all in marathi), Sunaparant in Devanagari- script Konkani. All are dailies. Other publications in the state include Goa Today (English language, monthly), Goan Observer (English weekly), Vavraddeancho Ixtt (Roman script Konkani weekly), Goa Messenger, Gulab (Konkani, monthly), Bimb (Devanagri script Konkani), Harbour Times, Digital Goa, and "J's House"

Chapter-5
Fishery Resources of Goa

Background

Fishing is one of the extractive occupations of mankind older than agriculture. Fish assumes greater significance to the people of goa as it forms the one of the most important items of the food of more than 90% population. Fishing serves as a means of livelihood to a large number of goans populations. About 60% of the total active fisherman are engaged in marine fishing and rest 40% in inland fishing. Altogether more than 5% of total working population of goa engaged in fishing activities. Apart from tourism, mining and manufacturing, fishing industry forms the second largest industry both in terms of employment and income. In addition to fishing the number of ancillary and subsidiary activities have grown around fish harvesting.

The Region Under Study

Location

Goa is a tiny emerald land situated well on the west coast of Indian peninsular. Goa is located between latitudes 15° N 48° 00 to 14° 53′ 54″ N and longitude 74° 20′ 13″ E to 73° 40′ 33″. It has 3702 sq km experiencing tropical oceanic climatic condition with profound orographic influence. Accordingly climate is balanced and moist through the year, supporting a total population of 13, 47,668 with average literacy rate is 82.01 as per 2001 census.

There are various important factors influencing development of fishing activities and resources of goa i.e. favorable coastline, balance climate. Good demand for fish and fishery products. Ready markets availability of finance excellent transport network cold storage facilities training institute for fisherman and good governance by the Directorate of Fisheries government of goa and other agencies.

FISHERY RESOURCES BASE

Goa consists of mainly two kinds of fisheries such as

1. Inland fisheries
2. Marine fisheries

Inland fisheries of goa are one of the riches source spread over 250 km in length. Inland fisheries are divided in two types i.e. Brackish and Fresh water fisheries.

Brackish eater fisheries includes extensive estuaries or river mouth, a large number of lagoons back waters and brackish eater lakes etc.

Fresh water fisheries are constituted by great river system, fresh water lakes vast networking irrigation canals, tanks resources and ponds etc. according to their system of management the inland fisheries have been further classified in to two types i.e. captured and cultivated.

Goa has a coastline of maximum 104 kms. It is a broken coastline characterized by numerous bays and head lands, groups of oceanic island with numerable creeks, mangroves swamps and coral reefs are believed to be extensive

TECHNIQUES

Several techniques are used in inland and marine fishing activities. Fishing is carried out in the river creeks of the four major rivers i.e. river Mandovi, Zuari, Chapora, Sal and Talpona with the help of gill nets, cast nets, barrier nets, besides this same fishing activities are taken up in the drains of Khazan lands i.e. Prawn farming fish culture etc. fishing activity is practiced with the help of stake nets in the river creeks through the year exceptionally high tides during the monsoon.

There are various techniques adapted for marine fisheries. Widely developed along the coasts as well as in deep seas i.e. operation at beach scenes gill nets fishing in offshore waters, cast net fishing in shallow waters, seasonal hook and line fishing and mechanised trawlers in deep seas.

Development of Fisheries in Goa

Goa has 250km of long network of inland waterways. Among the seven rivers/creeks originating from the Western Ghats and meeting the Arabian Sea, the Mandovi and Zuari are the medium size with Chapora, Sal, Terekhol and Galgibag of smaller size in order of importance. Around forty seven villages are involved in inland fishing. Besides being sanctuaries for the estuarine fisheries the rivers afford potential 12,000 hectares of Khazan lands along the banks of which around 3,500 hectares which is totally marshy can b gainfully developed for prawn faring/ pisciculture. There is around 400 tidal shellfish/ prawn filtration farms along the banks/ creeks and rivulets bordering t he rivers in the tidal zone. Around 199 hectares area has been developed for prawns farming during the last 5 years. This yield of prawns varies from 1 to 2 tonnes per hectare per crop. Over major parts of the estuaries in the tidal zone, traditional fishing is carried out. Such fishing is permitted under special legislation under the Daman and Dui fisheries rules 1981 with clearance of ports office from navigation point of view.

The fresh water resources of goa are comparatively very limited in the form of 3300 hectares spread in area of Selaulim reservoir in South Goa and 250 hectares at Anjunem dam in North Goa a notable fresh water lake in goa is the 7 hectares size Mayem Lake in Bicholim taluka.

FRESH WATER FISHERY RESOURCE

The state has 3300 ha of water spreads area of fresh water bodies covering Selaulim reservoir in south Goa and Anjunem Dam reservoir in North Goa. Besides, small water bodies are available in several irrigation bands in Mayem Lake and smaller perennial and seasonal water sheets. In this regards, fresh water aquaculture offers a good potential to increase fresh water fish production as well as to generate self employment to the inland rural youth.

Table below shows the area of fresh water fishery resources available in Goa

Sr. no	States/ U.Ts	Rivers and canals(kms)	Reservoirs (lakh Ha)	Tanks and ponds (lakh Ha	Flood plain lakes and derelict water (lakh Ha)	Blackish water (lakh Ha)	Total water bodies (lakh Ha)
	Goa	210.00	0.60	0.015	0.0016	0.0033	0.62
	Total	210.00	0.60	0.015	0.0016	0.0033	0.62

Source: Directorate of Fisheries (2008)

Culture fisheries

Number of prawn hatcheries	1
Number of licensed prawn farm	60
Area covered under science farm	176.31 ha
Area covered under traditional farming	600ha
Number of fresh water fish seed hatchery	1
Number of fresh water bodies	283
Total fresh water area	3962.00
Approximate products for culture fisheries	3526 tones
Demonstration farm	1

Source: Directorate of Fisheries (2008)

The production of inland fisheries has increased gradually over 30 years. Inland fish production was 1,508 tones in 1972, rose to 1959 in the year 1978, and it declined to 1328 tones in 1983 and rose again to 1585 in 1988. Within the span of 20 years t eh fresh water fish production doubles. In 1993, the production increased sharply to 3053 tones and in 1998 it further went up to 4283 tones. In the year 2005 the production reached to a remarkable figure of 4194 tones. The mentioned figures show a gradual and considerable growth in inland fish production.

IMPORTANT VARIETY WISE QUANTITY OF INLAND FISH CATCH (IN M. TONES)

Sr. no	Name of fish	2002	2003	2004	2005	2006	2007
1.	Prawn:						
	big	20	62	64	46	48	15
	medium	262	137	150	111	158	77
	small	958	977	987	716	979	811
2.	Lady fish	77	70	74	80	76	38
3.	Mullets	144	187	186	207	196	233
4.	Gerres		59	67	65	80	70
5.	Lutranus	28	23	35	45	37	05
6.	Cat fish	254	219	205	210	196	111
7.	Anchovy	31	22	31	34	35	03
8.	Pearl spot	59	62	59	71	64	116
9.	Betki	6	4	5	7	4	01
10.	Milk fish	-	-	-	1	-	30

Sr. no	Name of fish	2002	2003	2004	2005	2006	2007
11.	Mega lops	1	1	1	1	-	04
12.	Scatophagus	31	26	31	31	33	36
13.	Ambasis	111	94	83	92	88	09
14.	Crabs	157	133	143	161	138	116
15.	Black water clamps	60	723	756	716	718	604
16.	False clamps	293	194	179	163	155	70
17.	Oysters	19	81	380	184	167	1
18.	Balle	5	-	-	-	04	-
19.	Green clamps	46	613	364	464	358	88
20.	Miscellaneous	524	583	594	765	607	625
	Total	3147	4284	4397	4194	4131	3070

Source: Directorate of Fisheries, Govt of Goa. (2008)

MARINE FISHING

Marine fishing is a seasonal industry for a period of nine months. The fishing season commences in August soon after fishing ban is lifted and lasts till the end of may. Marine fisheries are practically closed during monsoon and fisherman utilize this period for mending their nits, maintenance of the fishing vessel and making new nets etc.

Marine fishery resources comprise of 104 kilometers of coastline characterized by innumerable creeks, bays, mangroves, swamps and coral reef, etc. Goas sea shelf extended up to 10,000 sq km 100 fathoms depth marine fisheries have developed extensively over the years in Goa due to favourable conditions. There are 6 talukas i.e. 42 villages along the coast engaged in extraction of development of marine fisheries. The highly productive fishable area in the sea extends up to 20-40 fathoms and covers approximately a total area of 2000 sq miles. In 1960, Goa exploited 20,000 tons of fish and there were hardly any mechanized trawlers to support 6 lakh human populations in the state.

Major part of the fishermen population of 42000 from the state is engaged in marine fisheries. In order to increase the fish production, mechanization has been introduced on a large scale and there are about 1157 mechanized fishing trawlers operating in the region. Besides there are 619 motorized country crafts, 109 FRP country are 859 non-motorized country crafts are engaged in fishing activities in Goa.

The following table shows the production important marine fisheries during 1960-2008

Year wise production figures of marine fisheries from 1960-2008

Year	Production in tones
1960	20,000
1963	17,000
1973	15,740
1983	35,165
1984	38,064
1985	39,926
1986	29,927
1988	40,713
1989	54,550
1990	53,179
1991	75,623
1992	96,333
1993	1,00,922
1994	95,840
1995	81,856
1996	92,737
1997	91,277
1998	67,236
1999	60,075
2000	64,563
2001	69,386
2002	67,563
2003	83,756
2004	84,395
2005	103,019
2006	96326
2007	91185
2008	88771

MARINE FISHERY RESOURCE POTENTIALITY- GOA

Sr. No	Talukas	Marine Fish Catching Centres	%	Population Marine Fishing Activites	%	Boats Mechanised	%	Boat Traditional	%	Prod In M Tones	%	Value In Rs (Crores)	%
1	Bardez	5	15.5	1085	6.89	195	31.7	165	19.43	15982	17.5	60.37	17.5
2	PERNEM	3	90	1193	758	71	11.5	61	7.1	8000	8.7	30.22	8.7
3	TISWADI	2	6.0	570	5.3	126	20.4	127	14.9	10083	11.0	38.08	11.0
4	SALCETE	6	18.1	5059	32.1	33	5.3	109	12.8	14595	16.0	55.13	16.0
5	MARMUGAO	9	27.2	5727	36.4	140	22.7	242	28.5	42383	46.4	160.06	46.4
6	CANACONA	8	24.2	2098	13.3	50	8.1	145	17.0	153	0.16	0.57	0.16
7	PONDA	-	-	-	-	-	-	-	-	-	-	-	-
8	QUEPEM	-	-	-	-	-	--	-	-	-	-	-	-
	TOTAL	33	100%	15732	100%	615	100%	849	100%	91185	100%	344.45	100%

SOURCE: DIRECTORATE OF FISHERIES. GOVERNMENT OF GOA 2007-2008.

FISHERY RESOURCE DEVELOPMENT

The main objectives of the present study is to assess the fishery resource development for the year 2007-08 based on the various attributes that have already discussed for specific point of time. The fishery resource development profile is calculated for the year 2007-08 as revealed in the table. The various attributes that have been selected for the index are:

1. Percentage share of marine fish catching centres in each taluka to the state total.
2. Percentage share of population involved in marine fishing activities in each taluka to the state total
3. Percentage share of mechanised boats used in fishing activities in each taluka to the state total.
4. Percentage share of traditional boats operated for fishing in each taluka to the state total.
5. Percentage share of production of marine fisheries in each taluka to the state total.
6. Percentage share of value of fisheries caught in each taluka to the state total.

The above mentioned six attributes have used to assess the level of fishery resource development to prepare a map by ranking co-efficient method on the basis of hierarchy of development. All talukas have been grouped in to five zones as noticed in the figure for 2007-08.

Name of the Talukas	% of Fish Catching Centres	Position	Population Engaged in Marine Fisheries	Position	% of Mechanised Boats	Position	% of Traditional Boats	Position	% of Production in Tones	Positon	% of Value of Fisheries in Crores	Position	Composite Indez (COI)
N. GOA													
BARDEZ	15.5	4	6.89	5	31.7	1	19.43	2	17.52	2	17.52	2	2.6
PERNEM	9	5	7.58	4	11.5	4	7.1	6	8.77	5	8.77	5	4.83
TISWADI	6	6	5.36	6	20.4	3	14.95	4	11.05	4	11.05	4	4.5
PONDA	-	-	-	-	-	-	-	-	-	-	-	-	-
S.GOA													
SALCETE	18.1	3	32.1	2	5.3	6	12.8	5	16	3	16	3	3.6
MORMUGAO	27.2	1	36.4	1	22.7	2	28.5	1	46.46	1	46.46	1	1.16
CANACONA	24.2	2	13.3	3	8.1	5	17	3	0.16	6	0.16	6	4.16

ZONE 1:

The first zone comprise of very high fishery resource development in the study area. It consists of one taluka such as Mormugao. This is the core area which indicates the maximum development of marine fishery resources. This zone is characterised by broken coast with numerous bay and headlands, creeks, swamps, coral reefs, etc. Situated well in the extreme west. Mormugao taluka has largest number of fisherman population which accounts 36.4% of total fisherman in the state engaged in fishing activities enthusiastically to earn their livelihood and make larger profits. This taluka has produced about 42373 tones valued at Rs. 160.06 crores in the year 2007-08. This zone also indicates the use of mechanized boats and traditional boats for fishing operations accounts for nearly 22.7% and 28.5% respectively. Excellent infrastructure facilities, banking, insurance, labour, ready markets are encouraging fishery activities and fishery resource development in this region. Therefore this zone can become a hub of fisheries in the study area. Further this activity stimulates the growth of ancillary and subsidiary economic activities to provide employment opportunities to the people. Thus this zone can become highly prosperous in the near future.

ZONE 2.

Bardez taluka is the second leading producer of marine fisheries enjoys suitable conditions for fishery resource development. This taluka has been placed in high zone. Bardez taluka is a coastal taluka with 9% of fish catching centres and 7.8% of total fisherman population engaged in fishing activities. It enjoys second largest positon in fishery resource development. It has first position in mechanised boats, second position in traditional boats, production, output values and fourth position in fish catching centres. Fifth position in fisherman population as well. However, this taluka has great potential for fishery resource development. With approximate measure fishery resource can be developed in this area on large scale thereby this zone could become highly progressive ensuring greater economic stability of fisherman community other people as well.

ZONE3:

Salcete taluka is the only taluka which falls in the moderate fishery resource development zone. It has 3rd place in fish catching centre production and output values, second position in fisherman population, fifth position in traditional boats and 6th position in the use of mechanised boats. Salcete taluka is rich in labour for fishing activities and favourable coast and other facilities encouraging the development for fishery activities in this zone.

ZONE4:

This zone is characterised by low fishery resource development. There are two talukas in this zone namely Tiswadi and Canacona taluka. Both the talukas are situated well along the coast. There is great scope for fishery resource development for the people in this zone. Comparatively these two talukas are lagging behind in respect of fishery resource production and output values. Tiswadi taluka has 4th position in production, output values and traditional boats, 3rd position in mechanised boats, and 6th position in percentage of fish catching centres and the number of population engaged in marine fishery activities.

However, the Canacona taluka records the lowest ranks 6th in output value and production, 3rd in traditional boats and fisherman population, 2nd in fish catching centres and 5th in mechanised boats. Thus there is enormous scope for development of fishery activities resources as well.

ZONE 5:

The last zone represents the very low fishery resource development. Pernem taluka lies in this zone. Pernem taluka enjoys 6th position in traditional boats, 5th position in fish catching centres, production, and output values, 4th position in fisherman population and mechanised boats. Fishery resource development is very limited due to various problems i.e. lack of proper infrastructure in this zone hence there is great scope for fishery resource development. People of Pernem taluka needs to be focussed on fishery activities by taking keen interest and avail the benefits of government as private agencies.

CONCLUSION

Fishery is an important extractive occupation of mankind older than agriculture. Fishery activity is also one of the major economic activities providing livelihood opportunities to large number of people and contributing 3% of GDP of the state. Near about 5% of total working population of Goa engaged in fishing activities. Fishery sector of Goa stimulating the growth ancillary and subsidiary activities which ensures greater economic stability of people in the state. Hence, development of fishery resources constitutes major part in the economic development of the state. It is hoped the present study of fishery resource potentialities at state level will be for other area of India.

If the state is to be made one of the prosperous in the country there is need for fishery resource development for this purpose few suggestions are given below:

1. We should retain the breeding grounds of fish and not destroy the nurseries of fish fauna
2. Patrolling boats and monitoring fleet are necessary to curb illegal fishing during ban period.
3. There should be allocation sufficient funds for infrastructure, cold storage facilities in under developed talukas.
4. Deep sea water pollution should be controlled by govt agencies.
5. Awareness needs to be created among fisherman community by organising seminar, conference and training programmes related to fishery resource development.
6. Fish catching centres in Canacona taluka need to properly utilize for increasing fishery resource.

Bibliography

A.G. Untawale (2004) : Know our Shore Goa WWF for Nature India Goa State, Office Panaji.

B.S. Negi (1997) : Geography of Resources. Kedarnath Publication, Meerut.

Balbir Singh Negi (1977) : Human Geography. Kedarnath Ramnath, Meerut.

Bari Mulay and T.M. Patil Gujrathi (1988) : Commercial Geography.

Directorate of fisheries Govt. of India (1991) : Fisheries Development in Goa. Government Printing Press, Panaji, Goa .

Directorate of Planning and Statistics (1992) : Goa Gazetteer, Govt. Printing Press, Panaji, Goa.

Dixit K.R. (1976) : Geomorphic Features of the West Coast of India between Bombay and Goa. Geographical Review of India, 38/3, pp. 260-281.

Ecoforum (2000) : Fish Curry Rice: A Citizens Report on the State of the Goan Environment. Mapusa : The Other India Press.

Economic Survey (2005-06) : Government of Goa, Directorate of Planning, Statistics and Evaluation, Panaji.

F.A.O. (2008) : Technical Guidelines for Responsible Fisheries Series, 4,5,11,12.

Goa Tourist Directory (1982) : Directorate of Tourism Government of Goa, Panaji.

Goa Tourist Directory (1982) : Directorate of Tourism, Government of Goa, Panaji.

Goh Cheng Leong and Gillian C. Morgan (1985) : Human and Economic Geography. Oxford University Press.

GoI (2001) : Agricultural Statistics at a Glance, 2001, Director of Economics and Statistics, Ministry of Agriculture, Government of India.

Gomantak Daily: 24 June 2010.

Government of Goa (1897) : The Indian Fisheries Act and The Goa Fisheries Rules. Govt. Printing Press, Panaji, Goa.

Government of Goa (2000) : Statistical Hand Book of Goa. Publication Division, Directorate of Planning, Statistics and Evaluation Panaji, Goa.

Government of Goa (2008) : Citizens' Charter for Captain of Ports and River Navigation Department, Govt. of Goa.

Government of Goa (2009) : Goa Economy in Figures. Directorate of Planning, Statistics and Evaluation, Panaji, Goa.

Govt. of Goa (2000) : The Indian Fisheries ACT 1987 and the Goa Fisheries Rules 1981. Govt. Printing Press.

Iyear S.D. and Wagle B.G. (1987) : Morphometric Analysis of the River Basin in Goa. Geographical Review of India, 49/2, pp. 11-18.

Joe D'Souza (2002) : Fishing Woes to the Fore. Goa Today.

Joseph K.M. (1985) : Marine Fishery Resources of India. In a System Framework of Marine Food Industry in India (Ed. E.R. Kulkarni and U.K. Srivastava), p. 90.

K.K. Khanna and Dr. V.K. Gupta (1994) : Economic and Commercial Geography. Sultan Chand and Sons Publishing.

K.N. Mohanta, S. Subramanian, N. Komarpant and A.V. Nirmale (2008) : Breeding of Gold Fish. Fishery Science Section ICAR, Ela, Old Goa. Sahyadri Offset System, Corlim, Goa.

Kamat Nandkumar (2003) : Threat to Goa's Khazan Farms' in Navhind Times, Dated: December 15.

M.S. Rao (1988) : Anmol's Dictionary of Geography. Anmol Publications, New Delhi.

Mineral and Export Intelligence (2009) : Goa Mineral Ore Exporters' Association, Goa.

Mrs. Usha Dessai (2006) : "Selection of Varieties of Prawns and Fish for Brackish Water Fish Farming".

Mrs. Usha Dessai and Mr. N.V. Verlekar : "Fish and Prawn Seed Resources of Goa". Directorate of Fisheries, Government of Goa.

Mulimani A.A. (1992) : The Geographical Analysis of Mineral Resource base Belgaum District for Development. M.Phil. Dissertation submitted to the Karnataka University, Dharwad.

N.P.S. Varde (2003) : Research Paper "Management of Coastal and Marine Environment".

Prabhakar S. Angle (1983) : Goa – An Economic Review. The Goa Hindu Association, Kala Vibhag.

Prabhakar S. Angle (1994) : Goa Concepts Mis-concepts. The Goa Hindu Association, printed by Arun Naik, Akashar Pratiroop Pvt. Ltd., Wadala, Bombay.

Prakash Shinde (1992) : Commercial Geography. Sheth Publishers Pvt. Ltd., Educational Publishers "Pallavi-Kunj" Margao, Goa.

Prakash Shinde (2008) : Geography of Resources. Sheth Publishers Pvt. Ltd., Bombay.

Prithwish Roy (1992) : Economic Geography – A Study of Resources. Amitabha Sen, New Central Book Agency (P) Ltd., Kolkata.

R.B. Mandal (1988) : Systems of Rural Settlements in Developing Countries. Concept Publishing Company, New Delhi.

Raikar A.V. (2003) : Research Paper "Sustainable Development of Marine Resources in Goa – What needs to be done?

Rashmi J. Desai (2007) : Geography of Resources. Vipual Prakashan, Mumbai.

Ravindra K. Majumdar (2006) : Value Added Fish Products. Fishing Chimes, Vol. 26, No. 4.

Rekha R. Gaonkar and R.B. Patil and Maria D.C. Rodrigues (2006) : Fishes and Fisheries. A.P.H. Publishing Corporation, 5 Ansari Road, Darya Ganj, New Delhi.

Shamila Manteiro (2003) : Research Paper "Fisheries Resources of Goa : Its Management and Conservation".

Shamila Monteiro (2009) : Farming of Catla and Rohu in Brackish Water Ponds: A Success Story. Fishing Chimes, July, Vol. 29, No. 4.

State of Goa (2008) : Indicators of Socio-Economic Development. Directorate of Planning, Statistics and Evaluation, Government of Goa, Panaji, Goa.

Statistics of Marine Product Exports (2005) : The Marine Products Export Development Authority. Ministry of Commerce and Industry, Govt. of India.

Z.A. Ansari and S.G. Dalai : "Over Exploitation of Fishery Resources, with Particular Reference to Goa". Directorate of Fisheries, Government of Goa.

Z.A. Ansari and S.G. Dalal and R.A. Sreepada (2003) : Research Paper "Ecosystem based Fishery Management for Sustainable Coastal Fisheries in Goa West Coast of India. National Institute of Oceanography, Dona Paula, Goa, India.

www.ingramcontent.com/pod-product-compliance
Ingram Content Group UK Ltd.
Pitfield, Milton Keynes, MK11 3LW, UK
UKHW021956270726
14060UKWH00002B/537